SPACE MATHEMATICS
Math Problems Based on Space Science

Bernice Kastner

DOVER PUBLICATIONS, INC.
Mineola, New York

Bibliographical Note

Space Mathematics: Math Problems Based on Space Science, first published by Dover Publications in 2012, is an unabridged republication of *Space Mathematics: A Resource for Secondary School Teachers,* originally published by NASA, Washington, D.C., in 1985.

International Standard Book Number
ISBN-13: 978-0-486-49033-5
ISBN-10: 0-486-49033-5

Manufactured in the United States by Courier Corporation
49033504 2015
www.doverpublications.com

TABLE OF CONTENTS

INTRODUCTION

High School mathematics teachers have long been aware that their students should know not only something about the development of pure mathematics but also something about its applications. Several years ago NASA, recognizing the appeal of aerospace activities, initiated and supported the development of curriculum supplements for several high school courses. Because attainments in aerospace would not be possible without mathematics, it was most appropriate that a supplementary publication dealing with space activities be prepared for teachers of mathematics.

The first mathematics curriculum supplement, *Space Mathematics, A Resource for Teachers,* was published in 1972. One of the most popular and oft-requested of the supplements, the book has been unavailable for several years. This volume updates the earlier work. We hope that a new generation of students will become more interested in mathematics as the result of seeing some of its significant applications in recent and current space projects. Working problems such as those in this book should enhance both the mathematical knowledge and skills of students and their appreciation and understanding of aerospace technology and achievements.

NASA's Technical Monitor for this project was Muriel M. Thorne, Educational Programs Officer, under the general direction of William D. Nixon, Chief of Education Services, NASA.

National Aeronautics and Space Administration
Washington, D.C.
September 1985

PREFACE

In 1972, a collection of mathematical problems related to space science entitled *Space Mathematics, A Resource for Teachers* was published by the Educational Programs Division of the National Aeronautics and Space Administration (NASA). As an early user of that publication, I can say that it has been both a pleasure and a challenge for me to undertake the revision of that volume of enrichment materials, especially in the light of another twelve years of activity in space exploration. This interval has been a period of much progress in both the science and the technology associated with the space program, and it has offered a wealth of new material in which to find applications of high school mathematics.

The basic format of the original publication has been retained, as well as many of the classical problems and those which complemented the new material. In developing the examples and problems presented here, we have aimed at preserving the authenticity and significance of the original setting while keeping the level of mathematics within the secondary school curriculum. The problems have been grouped into chapters according to the predominant mathematical topic. Within each chapter we have attempted, as far as possible, to group problems involving similar themes. There is a wide range of sophistication required to solve the various problems. Since this is a resource book for teachers, we have assumed that the reader will be interested not only in problems that can be brought directly into the classroom, but also in those that, although beyond the current level of their students, will increase the teacher's own awareness of some of the interesting applications of mathematics in the space program.

Perhaps the most valuable potential of a collection such as this lies in its ability to convey a sense of how secondary school mathematics is actually used by practicing scientists and engineers. Attitudes and approaches may thereby be fostered, on the part of teachers, that can help students to be more insightful users of the mathematics they learn. The present school mathematics curriculum, for example, gives no hint that many real-world problems do not have analytic solutions in closed form but may nevertheless be satisfactorily "solved" by using carefully chosen approximations or the numerical methods made possible by modern computers.

In this connection, we stress that in order to use numerical analysis correctly or to make good approximations, it is necessary to know something of the theoretical background of the subject and to understand the concepts of precision and accuracy and the use of significant digits. Also, methods that reveal meaningful aspects of a procedure are preferable to purely algorithmic prescriptions; the perhaps unfamiliar "factor unit" method of unit conversion presented in Chapter 2 is actually quite commonly used in science and engineering. It not only removes all uncertainty about whether to multiply or divide by a conversion factor but also is far more likely to contribute to an understanding of the underlying concepts than, for example, the more usual metric system algorithm expressed in terms of "moving" the decimal point.

Many NASA staff members contributed time and thought to this project, including personnel at the Goddard Space Flight Center, the Marshall Space Flight Center, the Jet Propulsion Laboratory, the Langley Research Center, and NASA Headquarters in Washington, D.C. These people, too numerous to mention individually, provided enthusiastic support, which is gratefully acknowledged.

Project Associate James T. Fey, of the University of Maryland, and reviewers Louise Routledge, Father Stanley J. Bezuszka, Gary G. Bitter, and Terry E. Parks, of the National Council of Teachers of Mathematics (NCTM) provided valuable comments and suggestions.

On the editorial and support side, I would like to thank the staff at the NCTM Reston office and Muriel M. Thorne, Educational Programs Officer at NASA Headquarters.

Bernice Kastner
Simon Fraser University
September 1985

CHAPTER ONE

MATHEMATICAL ASPECTS OF SOME RECENT NASA MISSIONS

Launch of the Space Shuttle *Challenger* on June 18, 1983.

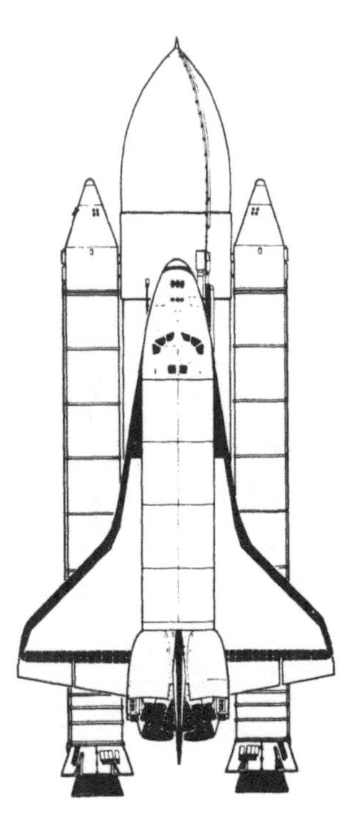

Fig. 1.1

Teachers of mathematics, like most adults in today's world, can hardly fail to be aware of the rapid development of space science. We realize that the spectacular achievements of the space program have depended heavily on mathematics—mathematics that is generally complex, advanced, and well beyond the level of most secondary school curricula. Even though this perception is valid, there are many significant aspects of space science that can be understood using only high school mathematics.

The exploration of space naturally uses the tools and techniques of astronomy. Astronomy in turn is gaining much new information as a result of sending scientific probes and satellites beyond Earth's atmosphere. Because astronomy has stimulated the growth of many of the concepts and methods of mathematics, the high school teacher will find here much that is familiar. However, in some instances the way mathematics is used to solve real-life problems is rather different from methods emphasized in school courses.

In this opening chapter, we shall examine several recent achievements of the National Aeronautics and Space Administration and identify mathematical ideas and questions that may be of interest to high school teachers and students. When appropriate, we will refer to a problem illustrating some aspect of the subject and worked elsewhere in the book.

The Space Shuttle

The Space Shuttle (Fig. 1.1) is a true aerospace vehicle—it takes off like a rocket, operates in orbit as a spacecraft, and lands like an airplane. To do this takes a complex configuration of three main elements: the Orbiter, a delta-winged spacecraft-aircraft, about the length of a twin-jet commercial airliner but much bulkier; a dirigible-like external tank, the only expendable element, secured to the Orbiter's belly and containing two million liters of propellant (Chapter 4, Problem 5); a pair of reusable solid rocket boosters, each longer and thicker than a railway tank car and attached to the sides of the external tank.

Each Space Shuttle is meant to be just one element in a total transportation system linking Earth with space. In addition to providing for continued scientific investigations by transporting such systems as the Spacelab and the Large Space Telescope, recently renamed the Edwin P. Hubble Space Telescope, into orbit (Chapter 3, Problem 4), the Space Shuttles are also expected to carry the building blocks for large solar-power space stations or huge antenna-bearing structures for improved communication systems (Chapter 4, Problems 9 and 10). Structures that would be too fragile to stand up under their own weight on Earth will be folded up in the Shuttle's cargo bay and assume their final shapes in the microgravity environment of space. The Shuttle will also be capable of carrying a work force of seven people and returning them home after the completion of their work.

One of the most basic mathematical problems raised by the launching and controlling of a Shuttle or any other spacecraft is that of describing its motion. This problem requires the ability to specify the position of the spacecraft's center of mass and its attitude (orientation) and to describe changes in both during flight. The specification of position and attitude can be accomplished by setting up suitable coordinate systems (Chapter 7, Problem 10). Instruments to determine a spacecraft's attitude are most effectively referenced to a spacecraft-based coordinate system, whereas ground control is best accomplished in terms of an Earth-based system. This dual-based system necessitates transformations between coordinate systems (Chapter 7, Problem 1, and Chapter 8, Problem 2). Describing a change of position and attitude requires an understanding of the measurement of time (Chapter 2, Problem 11). It is interesting to note here that our definition of a day on our rotating Earth must be redefined for a Space Shuttle Orbiter crew. For them the Sun might rise again and again every hour and a half!

The Planetary Probes

The launch of the two *Voyager* spacecraft in the summer of 1977 climaxed a series of fruitful missions of planetary exploration including the *Mariner*, *Viking*, and *Pioneer* series of probes to Mercury, Venus, Mars, Jupiter, and Saturn. All these missions sent back new information about the structure and composition of these planets and their associated moons. We focus in this book on some of the results of *Voyager 1* and *Voyager 2*. These probes, which benefited from more highly developed instrumentation and computer capability than their predecessors, approached closer to Jupiter (Chapter 7, Problem 11) and Saturn than previous flights did. Stunning pictures resulted, showing the unanticipated presence of active volcanoes on Jupiter's moon Io (Chapter 10, Problem 6) and the fine structure of Saturn's rings.

Among the mathematical problems that arose in these missions were the following.

1. Transmitting spacecraft observations back to Earth (Chapter 5, Problems 2 and 3, and Chapter 8, Problem 1).

2. Determining the time of transmission of spacecraft observations (Chapter 3, Problem 5).

3. Calculating the rotation period for planets such as Saturn, which is not solid and has no outstanding observable features like Jupiter's Great Red Spot (Chapter 2, Problem 13).

Satellites

NASA began its formal existence in 1958 and by the end of 1979 had successfully launched more than 300 large and small satellites with missions as diverse as observing Earth's weather (*Synchronous Meteorological Satellite* [SMS] series) and resources (*Landsat* series), providing communication links for television signals (*Applications Technology Satellite* [ATS] series), and measuring solar radiation outside Earth's atmosphere (*Orbiting Solar Observatory* [OSO] series).

The design of these satellites and their experiments and the analysis of the data gathered involve a variety of mathematical questions. We shall consider some of the following examples.

1. The connection between the conic sections and the law of gravitation (See Appendix).

2. For elliptic orbits, the connection between the orbit parameters and the period of revolution (Chapter 9, Problem 11) and the determination of the exact position of a satellite in its orbit at a specified time (Chapter 9, Problems 19 and 20).

3. The geometry necessary to correct for distortions arising when flat pictures are made of a curved Earth (Chapter 7, Problems 7 and 9, and Chapter 10, Problem 2).

4. The need for logarithms to understand how radiation is absorbed by Earth's atmosphere (Chapter 6, Problem 3).

5. The mathematical analysis of the reflective properties of the conic sections needed to design an X-ray telescope (Chapter 9, Problems 21 and 22).

6. The judicious use of approximation (Chapter 3, Problem 8; Chapter 4, Problems 6 and 8; Chapter 7, Problem 6; Chapter 9, Problem 22).

CHAPTER TWO

COMPUTATION AND MEASUREMENT

A photograph of the planet Jupiter made from images obtained by *Voyager 1* on February 5, 1979, showing the Great Red Spot and three of Jupiter's four largest satellites: Io (in front of Jupiter), Europa (brightly lit, to the right), and Callisto (barely visible at the bottom left).

Space science is based on a mathematical description of the universe. This mathematical description is in turn based on defining physical quantities clearly and precisely so that all observers can agree on any measurement of these quantities. Every measurement has two parts: a number and a unit. In mathematics, we tend to focus on the numbers and assume that the units are taken care of; but in scientific work, units receive careful attention through a procedure known as dimensional analysis, which is illustrated in the first problem.

Among the physical quantities used to describe the universe, some are considered fundamental quantities whereas others are derived quantities, comparable to the designation of definitions and undefined terms in a mathematical system. Although it does not really matter which particular quantities are the ones designated as fundamental, the most common are length, mass, and time. In scientific work the two major systems of units for these quantities are the mks (meter-kilogram-second) and the cgs (centimeter-gram-second). Every measurement is a comparison with the standards that are universally accepted as definitions of these fundamental units. In astronomy and space science, where large distances are common, the meter and even the kilometer are too small to be convenient; in Problems 5, 9, and 10 of this chapter, we show how more suitable units for length are defined.

Dimensional analysis (manipulation of units according to the rules of algebra) is the procedure used to ensure consistency in the definition and use of units. For example, since force is, by definition, the product of mass and acceleration, measured respectively in kg and m/s^2 in the mks system, the unit of force in this system must be equivalent to $kg \cdot m/s^2$. A new term, the *newton*, was created to describe the unit of force: 1 newton = $1\ kg \cdot m/s^2$.

PROBLEM 1. Newton's law of gravitation, one of the most important ideas in space science, states that the force of gravitational attraction between two bodies of masses M_1 and M_2 is proportional to the product of the two masses and inversely proportional to the square of the distance R separating the two masses. If G is the constant of proportionality, called the *universal gravitational constant*, this law can be stated in symbols as $F = \frac{GM_1M_2}{R^2}$. What must be the unit for G in the mks system?

Solution: Using dimensional analysis, we equate the known units in accordance with the relationship above without worrying about the numbers, then solve algebraically to get the unknown unit. This gives

$$\begin{aligned} \text{newton} &= (\text{unit for } G)\,(\text{kg})\,(\text{kg})\,(\text{m})^{-2} \\ \text{or (unit for } G) &= (\text{newton})\,(\text{m})^2\,(\text{kg})^{-2} \\ &= (\text{kg m sec}^{-2})\,(\text{m})^2\,(\text{kg})^{-2} \\ &= \text{m}^3\ \text{kg}^{-1}\ \text{sec}^2. \end{aligned}$$

PROBLEM 2. We know that in a circle of radius r, if an arc of length s subtends an angle θ and θ is measured in radians, then $s = r\theta$. Show that the radian is essentially dimensionless (i.e., an angle of $\pi/4$ radians is just the real number $\pi/4$).

Solution: Since r and s are both lengths, in the mks system they will both be measured in meters. From $s = r\theta$ we have $\theta = \frac{s}{r} = \frac{\text{m}}{\text{m}}$. Since the units cancel, θ is dimensionless.

Scientific theories and technological development both require accurate measurements. Since every measurement is an approximation, an important aspect of scientific and technical work is the analysis of experimental error and the control of the propagation of error when computations are made using measured quantities. The use of computers to solve complex problems by numerical methods has made error analysis even more important because computers approximate real numbers using finite decimals. Moreover, computers represent numbers internally using a floating point binary representation. Even though it is not really necessary to understand the binary numeration system to work with computers, such knowledge is essential to the analysis and control of error propagation in computational work. The next problem considers the floating point binary representation of our familiar numbers.

PROBLEM 3. The binary (i.e., base two) representation of a number uses only two digits, 0 and 1. Whereas in base ten the actual value of a digit is the product of its nominal value and the appropriate power of 10 according to the position of the digit with respect to the decimal point, in base two the value of a digit is the product of its nominal value and the appropriate power of 2. So, for example, the binary number 10011 has the value that we represent in base ten as $1 \times 2^4 + 0 \times 2^3 + 0 \times 2^2 + 1 \times 2^1 + 1 \times 2^0$, or $16 + 2 + 1 = 19$; the binary number 10.011 is the same as the decimal number $1 \times 2^1 + 0 \times 2^0 + 0 \times 2^{-1} + 1 \times 2^{-2} + 1 \times 2^{-3}$, or $2 + 0.25 + 0.125 = 2.375$.

a. Determine the binary representations of the decimal numbers 625, 6.25, and 0.0625.

Solution: 625 can be written as the sum

$$512 + 64 + 32 + 16 + 1 = 2^9 + 2^6 + 2^5 + 2^4 + 2^0,$$

$$\text{so } 625_{\text{ten}} = 1001110001_{\text{two}}.$$

6.25 can be written as the sum

$$4 + 2 + \frac{1}{4} = 2^2 + 2^1 + 2^{-2},$$

$$\text{so } 6.25_{\text{ten}} = 110.01_{\text{two}}.$$

$$0.0625 = \frac{625}{10000} = \frac{1}{16} = 2^{-4},$$

$$\text{so } 0.0625_{\text{ten}} = 0.0001_{\text{two}}.$$

b. Show that it is impossible to represent the decimal fraction 0.2 exactly in a finite binary code.

Solution: 0.2 = 2/10 = 1/5. To express this in binary notation, we must write 1/5 as a sum of unit fractions, each having some power of 2 as denominator. Since $1/2^3$ is the largest such fraction smaller than 1/5, we begin by finding the difference:

$$\frac{1}{5} - \frac{1}{2^3} = \frac{8-5}{5 \cdot 2^3} = \frac{3}{5 \cdot 2^3}$$

Now the largest unit fraction less than $3/(5 \cdot 2^3)$ with a power of 2 as denominator is $1/2^4$, so we next find the difference:

$$\frac{3}{5 \cdot 2^3} - \frac{1}{2^4} = \frac{6-5}{5 \cdot 2^4} = \frac{1}{5 \cdot 2^4}$$

This means that $1/5 = 1/2^3 + 1/2^4 + (1/2^4)(1/5)$. Since the fraction 1/5 has recurred, multiplied by $1/2^4$, we see that the first four digits we have found to the right of the binary "decimal point," 0.0011, will repeat continuously. In other words, $0.2_{\text{ten}} = 0.\overline{0011}_{\text{two}}$. (A quicker but less intuitive approach to finding this representation is to express 1/5 as the binary fraction 1/101 and then divide 101 into 1, using binary arithmetic.)

The reader can use the method of part (b) to show that the decimal fractions 0.1, 0.3, 0.4, 0.6, 0.7, 0.8, 0.9 also have infinitely repeating binary representations.

c. Almost all computers use a floating point binary representation for numbers. In this system, every number is expressed in the form $0.d_1d_2 \ldots d_n \times 2^m$, where $d_1 = 1$, $d_i = 0$ or 1 for i = 2, 3, . . . , n, and m is an integer. For example, the floating point representations of the numbers in part (a) would be

$$625 = 0.1001110001 \times 2^{10}$$

$$6.25 = 0.11001 \times 2^3$$

$$0.0625 = 0.1 \times 2^{-3}$$

Different computers have differing capabilities both with respect to the length (n) of the string of 0's and 1's that can be stored for any single number and with respect to the exponent m that can be stored. The limits available for n and m determine the largest and smallest number a computer can represent and also the size of the errors that must result when a number with an infinitely repeating representation must be stored with only a finite string length available.

If a certain computer can store only an eight-digit string ($n = 8$), then the representation for the decimal fraction 0.2 will be stored as 0.11001100×2^{-2}. What number is this, and what is the difference between this number and 0.2?

Solution:

$$0.11001100 \times 2^{-2} = \left(\frac{1}{2} + \frac{1}{4} + \frac{1}{32} + \frac{1}{64}\right) \times \left(\frac{1}{4}\right)$$

$$= \frac{1}{8} + \frac{1}{16} + \frac{1}{128} + \frac{1}{256}$$

$$= \frac{32 + 16 + 2 + 1}{256} = \frac{51}{256}$$

$$0.2 - \frac{51}{256} = \frac{1}{5} - \frac{51}{256}$$

$$= \frac{256 - 255}{5 \cdot 256} = \frac{1}{1280}$$

We now state two definitions used in error analysis. These definitions can be applied to both measurement errors and the errors that arise because of the way in which numbers are represented in computers. It is probably worth noting in this context that the term *measurement error* as used here does not imply that the measurement has been carelessly made but rather refers to the fact that every measuring instrument is limited in accuracy and can never provide more than an estimate of a true value.

Let X_T be the true value of a specified quantity, and let X be the value of this quantity as measured or as represented in the computer. Then:

$$\text{absolute error in } X = |X_T - X|$$

$$\text{relative error in } X = \left|\frac{X_T - X}{X_T}\right|$$

Observe that absolute error has the same units as the quantity under consideration, whereas relative error (usually reported as a percent) is dimensionless.

The relative error is considered to be the indicator of how good a measurement or any other approximation is. For example, a measurement of 2.5 mm with a possible absolute error of 0.05 mm has a relative error of $\frac{0.05}{2.5}$, or 2 percent, whereas a measurement of 1250 km, with a (much larger) possible absolute error of 5 km has a much smaller relative error of $\frac{5}{1250}$, or 0.4 percent. Awareness of the appropriate tolerance for relative error is a vital ingredient of scientific work.

PROBLEM 4. What are the absolute and relative errors if a computer that has an eight-bit binary digit string represents 0.2 as 0.11001100×2^{-2}?

Solution: From Problem 3c, the absolute error is 1/1280, or about 0.0008. The relative error $\doteq 0.0008/0.2 = 0.004$, or 0.4 percent.

The use of significant figures is helpful in error analysis. The number of significant figures is defined as the number of digits that can be assumed to be correct, starting at the left with the first nonzero digit, and proceeding to the right. By this definition, 10.62, 0.05713, and 4.600 all have four significant figures. A number such as 4300 is ambiguous. This ambiguity may be resolved by using scientific notation, since we may write the number as 4.3×10^3, 4.30×10^3, or 4.300×10^3 according to whether the number has two, three, or four significant figures, respectively.

When approximate numbers are added or subtracted, it can be shown that the absolute error in the sum or difference could be as large as the sum of the absolute errors of the individual numbers. When approximate numbers are multiplied or divided, it can be shown that the relative error of the result could be as large as the sum of the relative errors of the individual numbers. This means that for sums and differences of approximate numbers, the number of decimal places considered significant can never be greater than the number of decimal places in the least precise addend. For products and quotients, the number of significant figures can never be more than the smallest number of significant figures in the individual factors. Wherever appropriate, numerical results will be given in accordance with these guidelines.

PROBLEM 5. Earth's orbit around the Sun is elliptical, but in many cases it is sufficiently accurate to approximate the orbit with a circle of radius equal to the mean Earth-Sun distance of 1.49598×10^8 km. This distance is called the *Astronomical Unit* (*AU*). Listed in the chart that follows are actual Earth-Sun distances, given to five significant digits, on the first day of each month of a representative year. (The *American Ephemeris* lists daily distances and the actual times for these distances to seven significant digits.)

Date	*Distance* ($\times 10^8$ km)
1 January	1.4710
1 February	1.4741
1 March	1.4823
1 April	1.4949
1 May	1.5073
1 June	1.5169
1 July	1.5208
1 August	1.5183
1 September	1.5097
1 October	1.4977
1 November	1.4848
1 December	1.4751

a. To how many significant digits is it reasonable to approximate the Earth-Sun distance as though the orbit were circular?

Solution: To two significant digits, each of the distances in the table can be given as 1.5×10^8 km.

b. What are the largest possible absolute and relative errors in using the Astronomical Unit as the Earth-Sun distance in a computation instead of one of the distances from the table?

Solution:

$$(1.49598 - 1.4710) \times 10^8 = 0.0240 \times 10^8 \text{ km (smallest table value)}$$

$$(1.49598 - 1.5208) \times 10^8 = -0.0248 \times 10^8 \text{ km (largest table value)}$$

$$\text{absolute error} \leq 0.0248 \times 10^8 \text{ km}$$

$$\text{relative error} \leq \frac{0.0248}{1.49598} = 0.0166, \text{ or } 1.7 \text{ percent.}$$

The procedure of *dimensional analysis*, described earlier, is easily adapted and commonly used in science and technology for the task of unit conversion. Recall that in dimensional analysis the units are manipulated in accordance with the rules of algebra.

Suppose we wish to change a length of 623 cm to meters. The adaptation of dimensional analysis for unit coversion involves multiplication by a *factor unit* chosen according to the following simple principles: the factor unit is a fraction with a value of 1, whose numerator is expressed in terms of the unit we wish to have and whose denominator is expressed in terms of the unit we wish to change. Since 100 cm = 1 m, in order to change 623 cm to m, we perform the multiplication

$\frac{623 \text{ cm}}{1} \times \frac{1 \text{ m}}{100 \text{ cm}}$, "canceling" the cm in numerator and denominator to get

623/100 m, or 6.23 m.

More complex conversions can be done using multiplication by several factor units and those readers wishing to convert between British and metric units can also use this method. For example, the speed of light, 3.00×10^5 km/sec, can be found in miles per hour:

$$\frac{3.00 \times 10^5}{1} \frac{\text{km}}{\text{sec}} \times \frac{1 \text{ mile}}{1.61 \text{ km}} \times \frac{60 \text{ sec}}{1 \text{ min}} \times \frac{60 \text{ min}}{1 \text{ hour}} = 6.71 \times 10^8 \text{ miles per hour.}$$

PROBLEM 6. The deep space probe *Pioneer 10* took 21 months to get from Mars to Jupiter, a distance of 998 million kilometers. Use the factor unit technique to find its average speed in kilometers per hour during that period.

Solution:

$$\text{average speed} = \frac{\text{distance}}{\text{time}}$$

$$= \frac{998 \times 10^6 \text{ km}}{21 \text{ months}} \times \frac{12 \text{ months}}{365 \text{ days}} \times \frac{1 \text{ day}}{24 \text{ hours}}$$

$$= 6.5 \times 10^4 \text{ km/h, or about } 65\,000 \text{ km/h}$$

PROBLEM 7. **a.** Recall that the Astronomical Unit (Earth-Sun distance) discussed in Problem 5 is 1.496×10^8 km, to four significant figures. Find the Earth-Sun distance in miles to three significant figures.

Solution:

$$1 \text{ AU} = 1.496 \times 10^8 \text{ km}$$

$$= \frac{1.496 \times 10^8 \text{ km}}{1} \times \frac{1 \text{ mile}}{1.61 \text{ km}}$$

$$= 9.29 \times 10^7 \text{ miles (almost 93 million miles)}$$

b. The chart that follows gives the mean distance in kilometers of each planet in the solar system from the Sun. Express these distances in AU, using a suitable number of significant digits.

Planet	*Distance* (km × 10^8)
Mercury	0.579
Venus	1.08
Mars	2.27
Jupiter	7.78
Saturn	14.3
Uranus	28.7
Neptune	45.0
Pluto	59.1

Solution: Since each distance in the table has three significant digits, and the factor unit $\frac{1 \text{ AU}}{1.49598 \times 10^8 \text{ km}}$ has an exact number in the numerator and six significant digits in the denominator, the distances in AU can be given to three significant digits. Multiplying by the factor unit shown gives the following distances in AU:

Mercury	0.387
Venus	0.722
Mars	1.52
Jupiter	5.20
Saturn	9.56
Uranus	19.2
Neptune	30.1
Pluto	39.5

PROBLEM 8. The *Solar Maximum Mission* (SMM) satellite orbits Earth at a height of 560 km. In many computations, the Earth-Sun distance of 1.5×10^8 km is used to approximate the distance of SMM from the Sun. What is the maximum relative error of this approximation?

Solution: The distance of SMM from the Sun is contained within the range (Earth-Sun distance ± (Earth diameter + 560 km)), or (Earth-Sun distance ± 6930 km). From Problem 5, if 1.5×10^8 km is used as the Earth-Sun distance, the absolute error $\leq 2.48 \times 10^6$ km. 6930 km is much smaller than this error, so that the absolute and relative errors incurred in using 1.5×10^8 km as the SMM-Sun distance are the same as those of part (b) of Problem 5. If greater accuracy is required, it will be necessary to use daily *Ephemeris* values such as those in the listing given in Problem 7. In this event, it is still true that since 6930 km $< 7 \times 10^3$ km, the relative error in approximating the SMM-Sun distance with the Earth-Sun distance $< \frac{7 \times 10^3}{1.5 \times 10^8} = 4.67 \times 10^{-5}$, or about 0.005 percent.

The Astronomical Unit (AU), although useful for measuring distances within the solar system, is too small to be convenient for distances to stars. We shall therefore consider two other units of length used by astronomers. The first is called the *light-year.*

PROBLEM 9. The light-year is the distance traveled by light during one Earth year. To three significant digits, the speed of light is 3.00×10^5 km/s. Find the length of the light-year in km and in AU.

Solution: 1 Earth year = 365.25 days = $365.25 \times 24 \times 60 \times 60$ seconds. In one year, light travels $3.00 \times 10^5 \times 365.25 \times 24 \times 60 \times 60$ km $= 9.47 \times 10^{12}$ km. To express this distance in AU, 1 light-year $= 9.47 \times 10^{12}\text{ km} \times \frac{1\text{ AU}}{1.50 \times 10^8\text{ km}} = 6.31 \times 10^4$ AU.

The *parsec* is the astronomical unit of distance that relates to observational measurements. In order to define this unit, we must consider the fact that when we observe the heavens, we have no direct perception of depth or distance. A useful model developed to portray the heavens is the celestial sphere. In this model, Earth is surrounded by an imaginary sphere with infinite radius. A coordinate system, similar to latitude and longitude, is imposed on the celestial sphere by projecting Earth's rotation axis on the sphere to identify the celestial north pole (CNP) and celestial south pole (CSP) as shown in Fig. 2.1. Since the radius of the celestial sphere is infinite, all parallel lines point to the same spot on the sphere, and so every line parallel to Earth's rotation axis also points to the celestial north and south poles.

The extension of Earth's equatorial plane intersects the celestial sphere in a great circle called the celestial equator. Now a system of small circles of *declination* (δ), comparable to latitude circles on Earth, is imagined on the celestial sphere, and a system of great circles called *right ascension* (α) circles, comparable to longitude, passing through the two poles, completes the coordinate system (Fig. 2.2).

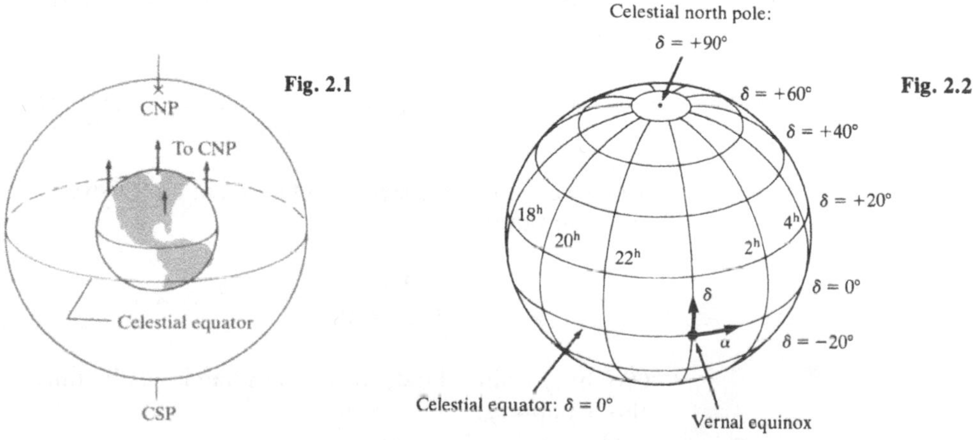

Fig. 2.1

Fig. 2.2

Every star or celestial object can now have its position identified by the ordered pair (α, δ). Because Earth rotates with respect to the celestial sphere, the time of observation must also be known in order to use the coordinate system. Differences in the positions of two objects on the celestial sphere are expressed in terms of the angle subtended at Earth by the arc joining these points.

As Earth revolves around the Sun, very distant stars show no discernible changes in position, but closer stars will show apparent motion with respect to the celestial sphere when viewed from different points in Earth's orbit, as shown in Fig. 2.3. This apparent motion is called *parallactic motion*, and the change in position is called the *parallax angle*. In this context, 1 parsec is defined as the distance at which the radius of Earth's orbit subtends an angle measuring 1 arc-second (see Fig. 2.4).

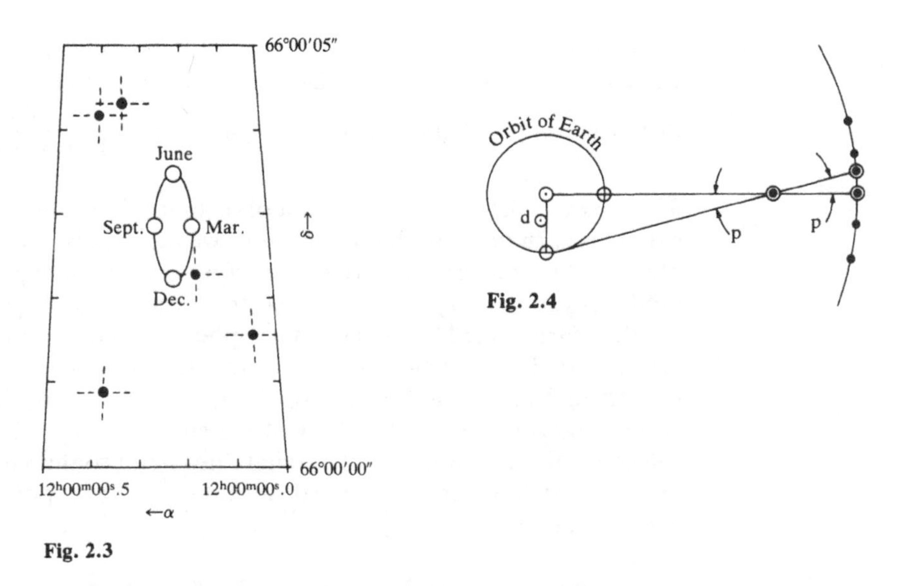

Fig. 2.3

Fig. 2.4

PROBLEM 10. **a.** Find the length to three significant digits of 1 parsec in terms of AU, km, and light-years.

Solution: If θ is in radians, we have arc length = $r\theta$, where r is the distance expressed in the same units as the arc length. In this case, arc length = radius of Earth's orbit = 1 AU.

$$\theta = 1 \text{ second} = \frac{1}{3600} \text{ degree} = \frac{1 \text{ degree}}{3600} \times \frac{\pi \text{ radians}}{180 \text{ degrees}}$$

Since we want three significant digits in our answer, let us use $\pi = 3.142$ in this computation:

$$\theta = \frac{3.142}{3600 \times 180} \text{ rad} = 4.85 \times 10^{-6}$$

(We have omitted rad, since the radian is really dimensionless. (Problem 2, this Chapter))

Then

$$r = \frac{\text{arc length}}{\theta} = \frac{1\ \text{AU}}{4.85 \times 10^{-6}} = 2.06 \times 10^5\ \text{AU}.$$

In terms of km, $r = 2.06 \times 10^5 \times 1.50 \times 10^8$ km $= 3.09 \times 10^{13}$ km. Since 1 light-year $= 6.31 \times 10^4$ AU, 1 parsec $= 2.06 \times 10^5\ \text{AU} \times \frac{1\ \text{light-year}}{6.31 \times 10^4\ \text{AU}} = 3.26$ light-years.

b. In general, if p is the parallax of a star and d its distance from Earth, then $d\ (\text{in parsecs}) = \frac{1}{p\ (\text{in seconds of arc})}$. The parallax of our nearest star, α Centauri, is 0.75 seconds, and the parallax of Sirius, one of the brightest stars in the northern sky, is 0.38 seconds. Find the distances to these stars in parsecs and in km.

Solution: For α Centauri,

$$d = \frac{1}{0.75}\ \text{parsecs} = 1.3\ \text{parsecs}$$

$$= 1.3 \times 3.09 \times 10^{13}\ \text{km}$$

$$= 4.1 \times 10^{13}\ \text{km}.$$

For Sirius,

$$d = \frac{1}{0.38}\ \text{parsecs} = 2.6\ \text{parsecs}$$

$$= 2.6 \times 3.09 \times 10^{13}\ \text{km}$$

$$= 8.1 \times 10^{13}\ \text{km}.$$

The accurate measurement of time has been one of the most challenging problems in human history. We now tend to take for granted the civil time-keeping system in general use. This system has evolved over many centuries and from time to time has been substantially revised. The original definitions of day, month, and year depended, respectively, on observations of the periodic motions of the Earth, Moon, and Sun with respect to the celestial background as observed from Earth. Since all these motions have fluctuations in their periods, it is not possible to define a completely regular unit on which to base an accurate time measurement in terms of the day, month, or year.

The first time-keeping instrument that did not depend on celestial observation was the pendulum clock. It did, however, depend on the Earth's gravity, which varies with geographic location and the positions of the Sun and Moon. The second, originally defined as $\frac{1}{24 \times 60 \times 60}$ of a day, more recently has been redefined in terms of the microwave emissions of certain atoms (e.g., quartz crystals). This new definition provides a uniform standard with which to measure intervals of time.

PROBLEM 11. **a.** One of the serious problems of the old nonuniform time units was the accumulation of error. It might seem that an accuracy of 1 second a day (a possible relative error of $\frac{1}{24 \times 60 \times 60}$, or about 1×10^{-5}) would be sufficient for most technical or scientific purposes. Show that an error of 1 second a day could result in an error of 1.1×10^4 km in the position of the Earth in its orbit after only 1 year. (Assume Earth's orbit is circular with a radius of 1.5×10^8 km.)

Solution: In one year the error could be 365 seconds. Earth moves through an angle of $\frac{2\pi}{365 \times 24 \times 60 \times 60}$ radians in one second, or $\frac{2\pi}{24 \times 60 \times 60}$ radians in 365 seconds. The length of arc that subtends this angle in a circle of radius 1.5×10^8 km is

$$s = r\theta = 1.5 \times 10^8 \times \frac{2\pi}{24 \times 60 \times 60} = 1.1 \times 10^4 \text{ km}.$$

b. The *tropical year* is defined as the time difference between successive vernal equinoxes—in other words, the time it takes Earth to complete one revolution around the Sun. This time does not have a simple relationship to Earth's rotation period (the day). In fact, it turns out that to the nearest second one tropical year is 365 days, 5 hours, 48 minutes, 46 seconds. Show that the current system of adding an extra day to each calendar year that is a multiple of 4 but not a multiple of 100 (leap years) serves to give each calendar year an integral number of days and also keeps the seasons constant with respect to the calendar.

Solution: If a calendar year has 365 days, the excess time in a tropical year is 5 h 48 m 46 s, not quite 1/4 day. Multiplying this excess by 4, $4 \times$ (5 h 48 m 46 s) = 23 h 15 m 4 s, almost 1 day. If we add 1 extra day each 4 years, we will create a deficit of 24 h − (23 h 15 m 4 s) = 44 m 56 s for each leap year. In each 100 years, there are 25 years that are multiples of 4; however, after 24 leap years, the deficit will accumulate to $24 \times$ (44 m 56 s) = 17 h 58 m 24 s, almost 3/4 day. This will almost balance the excess accumulation for the remaining 4 years of the century, so that years that are multiples of 100 should not be leap years. It is clear that further juggling will be necessary, since things never balance exactly.

c. For some computations in astronomy and space science, it is necessary to have an absolute time that is a continuous count of the number of time units from some arbitrary reference. The universally accepted standard is the Julian Day Calendar, a continuous count of the number of days since 12:00 noon on 1 January 4713 B.C. This curious starting date was actually chosen in A.D. 1582 by considering the cycle that is the least common multiple of the 28-year solar cycle (the interval required for all dates to recur on the same day of the week), the 19-year lunar cycle (the interval containing an integral number of lunar months), and the 15-year indiction (the tax period introduced by the Roman emperor Constantine in A.D. 313). The year 4713 B.C. was the most recent date prior to 1582 when these cycles coincided, and it had the added advantage of predating the ecclesiastically approved date of Creation, 4 October 4004 B.C. How long is the Julian day cycle, and when is the next year when all three of the cycles used in its creation will coincide?

Solution: The least common multiple of 28, 19, and 15 is their product, since these numbers have no prime factors in common. $28 \times 19 \times 15 = 7980$, so the next year the cycles coincide will be $(-4713) + 7980$, or A.D. 3267.

d. A clever computer algorithm for converting calendar dates to Julian days was developed using FORTRAN integer arithmetic (H. F. Fliegel and T. C. Van Flandern, "A Machine Algorithm for Processing Calendar Dates," *Communications of the ACM* 11 [1968]: 657). In FORTRAN integer arithmetic, multiplication and division are performed left to right in the order of occurrence, and the absolute value of each result is truncated to the next lower integer value after each operation, so that both 2/12 and −2/12 become 0. If I is the year, J the numeric order value of the month, and K the day of the month, then the algorithm is

$$\text{JD} = K - 32075 + 1461 * (I + 4800 + (J-14)/12)/4$$
$$+ 367 * (J-2-(J-14)/12*12)/12 - 3 * ((I + 4900 + (J-14)/12)/100)/4.$$

The calendar date 25 December 1981 is JD 2 444 964. Use a hand calculator and this algorithm to find the Julian dates of the launch of *Explorer 1* (the first U.S. satellite placed into orbit), Greenwich Mean Time 1 February 1958 (Eastern Standard Time January 31, 1958), and the launch of the seventh Space Shuttle on 18 June 1983 (carrying the first American female astronaut, Sally Ride, into orbit).

Solution: For 1 February 1958, $I = 1958, J = 2, K = 1$.

$$\text{JD} = 1 - 32075 + 1461*(1958+4800+(2-14)/12)/4$$
$$+ 367*(2-2-(2-14)/12*12)/12 - 3*((1958+4900+(2-14)/12)/100)/4$$
$$= 1 - 32075 + 1461*6757/4 + 367*(1*12)/12 - 3*(6857/100)/4$$
$$= 1 - 32075 + 2\,467\,994 + 367 - 51 = 2\,436\,236$$

For 18 June 1983, $I = 1983, J = 6, K = 18$.

$$\text{JD} = 18 - 32075 + 1461*(1983+4800+(6-14)/12)/4$$
$$+ 367*(6-2-(6-14)/12*12)/12 - 3*((1983+4900+(6-14)/12)/100/4$$
$$= 18 - 32075 + 1461*6783/4 + 367*4/12 - 3*68/4$$
$$= 18 - 32075 + 2\,477\,490 + 122 - 51 = 2\,445\,504.$$

A large number of satellites require ground processing of spacecraft sensor data to determine the spacecraft attitude (i.e., the spacecraft's orientation). Examples of sensors used are Sun sensors, Earth sensors, and star sensors. These sensors provide information, usually a measured angle, concerning the spacecraft *pointing* relative to a celestial body (e.g., Sun, Earth, or star).

Telemetry signals from these sensors are converted on the spacecraft to digital counts and transmitted to ground stations. The digital count representation of a sensor output can be easily converted to meaningful measurements and units on the ground. However, telemetry signals are frequently subject to random interference, or "noise." To understand the meaning of noise, one has only to tune into a weak channel on a television set; the "snow" that is seen is a visual representation of noise present in an electronic signal. Noise consists of random signals superimposed on valid electronic signals from any electronic device. An in-depth understanding of the cause, effect, and reduction of noise is not necessary in the context of this problem. However, it should be understood that noise can sufficiently corrupt any electronic signal to the extent that making use of, and properly interpreting, the true signal can be difficult.

This problem applies to spacecraft instruments and sensors. A number of methods have been developed to smooth data and remove the effects of noise. In the next problem, we examine one such method, called the running average.

PROBLEM 12. Given an ordered set of numbers, X_j, j = 1, 2, . . . M, a smoothed set of numbers can be found by averaging each number with the *n* preceding and the *n* following numbers. Symbolically,

$$X'_j = \frac{1}{2n + 1} \sum_{j-n}^{j+n} X_i,$$

where *n* is typically a small whole number ($n \leq 5$).

For example, for the data in Table 2.1, if $n = 2$, then

$$X'_7 = \frac{1}{5} (X_5 + X_6 + X_7 + X_8 + X_9)$$

$$= \frac{1}{5} (11 + 14 + 18 + 19 + 16) = 15.6.$$

a. Compute the smoothed values of the data in Table 2.1 for $n = 1$ and $n = 2$.

Solution: We present a computer program in BASIC, along with the run for this task.

Table 2.1
Spacecraft Sensor Data

Sample No. (j)	Unsmoothed Value (X_j)	Sample No. (j)	Unsmoothed Value (X_j)
1	2	11	20
2	7	12	20
3	10	13	18
4	6	14	19
5	11	15	20
6	14	16	20
7	18	17	17
8	19	18	19
9	16	19	18
10	17	20	16

```
10  REM  SMOOTHING FUNCTION
20  REM  ROUNDED TO 2 DECIMAL PLA
     CES
100  DATA  2,7,10,6,11,14,18,19,1
     6,17,20,20,18,19,20,20,17,19
     ,18,16
110 M = 20
120  DIM X(M): DIM Y(M): DIM Z(M)

130  FOR J = 1 TO M
140  READ X(J)
150  NEXT J
160  FOR J = 2 TO 19
170 Y(J) = (X(J - 1) + X(J) + X(J
      + 1)) / 3
180 Y% = 10 * (Y(J) + 0.05):Y(J) =
     Y% / 10
190  NEXT J
200  FOR J = 3 TO 18
210 Z(J) = (X(J - 2) + X(J - 1) +
     X(J) + X(J + 1) + X(J + 2)) /
     5
220 Z% = 10 * (Z(J) + 0.05):Z(J) =
     Z% / 10
230  NEXT J
240  PRINT "J";: HTAB 10: PRINT "
     X(J)";
245  HTAB 20: PRINT "X'(J),N=1";:
      HTAB 30: PRINT "X'(J),N=2"
250  PRINT "1";: HTAB 10: PRINT X
     (1)
260  PRINT "2";: HTAB 10: PRINT X
     (2);: HTAB 20: PRINT Y(2)
270  FOR J = 3 TO 18
280  PRINT J;: HTAB 10: PRINT X(J
     );: HTAB 20: PRINT Y(J);
285  HTAB 30: PRINT Z(J)
290  NEXT J
300  PRINT "19";: HTAB 10: PRINT
     X(19);: HTAB 20: PRINT Y(19)

310  PRINT "20";: HTAB 10: PRINT
     X(J)
320  END

]RUN
J         X(J)       X'(J),N=1  X'(J),N=2
1         2
2         7          6.3
3         10         7.7        7.2
4         6          9          9.6
5         11         10.3       11.8
6         14         14.3       13.6
7         18         17         15.6
8         19         17.7       16.8
9         16         17.3       18
10        17         17.7       18.4
11        20         19         18.2
12        20         19.3       18.8
13        18         19         19.4
14        19         19         19.4
15        20         19.7       18.8
16        20         19         19
17        17         18.7       18.8
18        19         18         18
19        18         17.7
20        18
```

b. If the data points are plotted and joined by line segments, we get a graph demonstrating data fluctuations. Normally, what is of interest is the underlying smooth curve (hence the term *smoothing*) for this process. Compare the graphs of the unsmoothed data and the smoothed data for $n = 2$.

Solution: Fig. 2.5 shows the plot of the unsmoothed data (solid line) and the data smoothed with $n = 2$ (broken line).

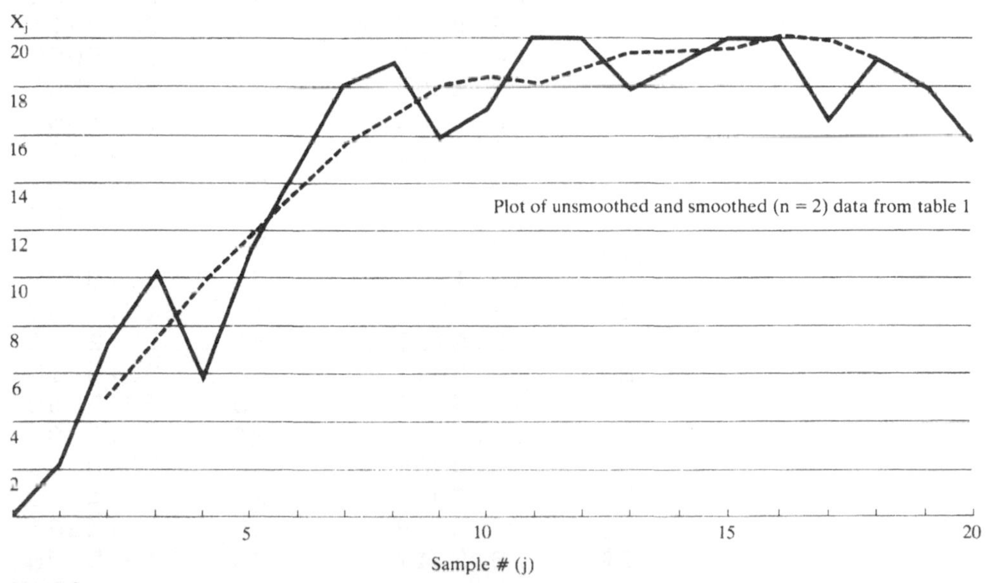

Fig. 2.5

c. This smoothing technique cannot be used blindly. It is possible to disguise the true nature of the data by smoothing. Modify the program of part (a) to smooth the data in Table 2.2 and compare the graphs of the unsmoothed data and the smoothed data for some value of n, for instance, 5. How has the smoothing technique disguised the true nature of the data?

Solution: The following program incorporates a subroutine to smooth the data of Table 2.2 for values of n from 1 through 5. The graph of the unsmoothed data and the smoothed values for $n = 5$ are displayed in Fig. 2.6. We see that the original data had a sinusoidal form, with the noise appearing as some slight departures from the smooth curve. The smoothed data are still sinusoidal, but the amplitude has been drastically reduced. It is evident, if the program with these data is run, that each increase in n reduces the amplitude more than the previous n.

Table 2.2

j	X_j	j	X_j	j	X_j
1	20.5836	35	25.7612	69	12.4995
2	24.4349	36	24.8147	70	10.5644
3	28.8846	37	18.7918	71	9.0069
4	27.1585	38	14.7649	72	9.1803
5	27.5732	39	10.7724	73	9.1452
6	24.4361	40	8.6446	74	14.1750
7	21.2117	41	7.0319	75	17.2637
8	14.6925	42	8.6086	76	21.4681
9	12.5582	43	11.4900	77	25.4627
10	8.0117	44	15.3886	78	27.0909
11	8.1619	45	18.4432	79	28.1451
12	9.0843	46	20.6217	80	25.6367
13	10.1741	47	24.7681	81	24.9373
14	15.2122	48	27.5421	82	20.6476
15	17.2274	49	28.1035	83	15.7963
16	20.9153	50	27.3259	84	12.2591
17	25.4725	51	23.4475	85	8.2834
18	26.1255	52	20.5059	86	9.4840
19	28.2650	53	16.3552	87	8.5730
20	26.0446	54	10.8868	88	10.2066
21	23.3059	55	10.0192	89	14.0093
22	19.1884	56	8.2502	90	18.7170
23	15.7242	57	9.5464	91	22.6785
24	12.9586	58	9.1521	92	25.6463
25	10.1285	59	13.9209	93	26.2984
26	6.2595	60	18.6782	94	27.7386
27	8.6186	61	23.2414	95	27.3292
28	9.5012	62	26.3716	96	23.7517
29	15.3659	63	28.5849	97	19.3016
30	18.2059	64	28.9297	98	15.3903
31	22.4038	65	25.8980	99	12.9370
32	25.5969	66	22.4440	100	10.0038
33	28.4748	67	21.3365		
34	29.2417	68	15.3698		

```
10  REM  SMOOTHING FUNCTION 2
20  REM  USING SUBROUTINE
100  DATA  20.5,24.4,28.9,27.2,27
     .6,24.4,21.2,14.7,12.6,8.0
101  DATA  8.2,9.1,10.2,15.2,17.2
     ,20.9,25.5,26.1,28.3,26.0
102  DATA  23.3,19.2,15.7,13.0,10
     .1,8.3,8.6,9.5,15.4,18.2
103  DATA  22.4,25.6,28.5,29.2,25
     .8,24.8,18.8,14.8,10.8,8.6
104  DATA  7.0,8.6,11.5,15.4,18.4
     ,20.6,24.8,27.5,28.1,27.3
105  DATA  23.4,20.5,16.4,10.9,10
     .0,8.3,9.5,9.2,13.9,18.7
106  DATA  23.2,26.4,28.5,28.9,25
     .9,22.4,21.3,15.4,12.4,10.6
107  DATA  9.0,9.2,9.1,14.2,17.3,
     21.5,25.5,27.1,28.1,25.6
108  DATA  24.9,20.6,15.8,12.3,8.
     3,9.5,8.6,10.2,14.0,18.7
109  DATA  22.7,25.6,26.3,27.7,27
     .3,23.8,19.3,15.4,12.9,10.0
110 M =  100
120  DIM X(M): DIM Y(M): DIM Z(M)

130  FOR J =  1 TO M
140  READ X(J)
150  NEXT J
160  FOR N =  1 TO 5
170  GOSUB 1000
180  NEXT N
190  END
1000  REM  SMOOTHING SUBROUTINE
1010  PRINT "J","X(J)";: HTAB 28:
      PRINT "X'(J),N=";N
1020  FOR I =  1 TO (N + 1)
1030  PRINT I,X(I)
1040  NEXT I
1050  FOR J =  (N + 1) TO (M - N)
1060 SUM =  0
1070  FOR I =  (J - N) TO (J + N)
1080 SUM =  SUM + X(I)
1090  NEXT I
1100 Y(J) =  SUM / (2 * N + 1)
1110 Y% =  100 * (Y(J) + 0.005):Y(
     J) =  Y% / 100
1120  PRINT J,X(J),Y(J)
1130  NEXT J
1140  FOR I =  (M - N + 1) TO M
1150  PRINT I,X(I)
1160  NEXT I
1170  RETURN
```

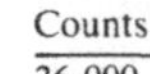

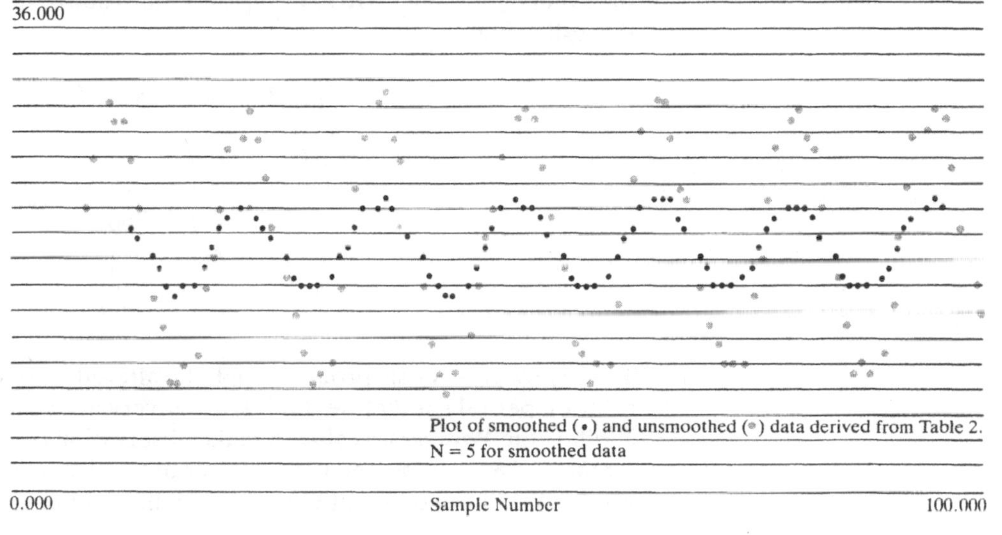

Fig. 2.6

We end this chapter with a problem that shows how a classical mathematical model is modified so that it can be used to determine the period of rotation of a planet. The modification uses some of the principles of scientific accuracy discussed earlier in this chapter and illustrates the use of successive iterations to refine a computation, easily done by computer.

The procedure we develop has been used to determine the rotation period of Saturn more accurately than earlier estimates by using observations of variations in the planet's radio emissions made by *Voyager 1*. Since this planet has neither a solid surface nor any distinctive atmospheric features comparable to Jupiter's Great Red Spot, what is computed is the period of rotation of the magnetic field of the planet. Because of the complex nature of the radio emission data, we illustrate the method by computing the rotation period of Jupiter rather than Saturn.

We begin with a question that is essentially the same as the familiar "How much time elapses between successive alignments of the hands of the clock?" but that sets the stage for the actual problem we wish to solve.

PROBLEM 13. **a.** Jupiter rotates on its axis once every 9.92 hours, and its moon Io revolves around Jupiter once every 42.5 hours. What is the length of time between consecutive passages of Io over a particular spot on Jupiter?

Solution: Let R_J and R_I be the angular rotation and revolution rates for Jupiter and Io respectively.

Then
$$R_J = \frac{360}{9.92} = 36.3 \text{ degrees/hour}$$

and
$$R_I = \frac{360}{42.5} = 8.47 \text{ degrees/hour.}$$

In Fig. 2.7, Io moves from A to B while the point S on Jupiter makes a complete revolution and then goes on to S′ to be under Io again. So we must find the time T such that $R_J T - 360 = R_I T$. Using the values above for R_J and R_I, we get the following:

$$36.3\,T - 360 = 8.47\,T$$

$$36.3\,T - 8.47\,T = 360$$

$$(27.8)\,T = 360$$

$$T = \frac{360}{27.8} = 12.9 \text{ hours}$$

To see how this classic problem might be altered, suppose we don't know Jupiter's rotation period (or that we don't know it very accurately). As it approached Jupiter, the *Voyager* was able to make observations of the times at which Jupiter's Great Red Spot appeared in the center of the disc as viewed from *Voyager*. We want to use these observations to determine Jupiter's rotation period.

b. *Voyager's* trajectory as it approached Jupiter is illustrated in Fig. 2.8. Modify the results of part (a) to find Jupiter's period if the Red Spot is observed to be in the center at time t_1 = 2 h 25 min ± 1 min, when *Voyager's* distance from Jupiter is $D_1 = 7.70 \times 10^5$ km, and again at time t_2 = 16 h 24 min ± 1 min, when *Voyager's* distance from Jupiter is $D_2 = 4.72 \times 10^5$ km and *Voyager* has moved through an angle $\alpha = 147.2°$ with respect to Jupiter's center between these two observations.

Solution: If *Voyager's* trajectory were circular and if Jupiter's period is P, then the analysis of part (a) can be applied with $R_J = 360/P$ and $R_I T = \alpha$. We now have $(360/P)T - 360 = \alpha$, which transforms into $T = P + \alpha P/360$. However, the trajectory is not circular, so we must take into account the different lengths of time it takes for light to travel from Jupiter to *Voyager*. Since light travels at a speed c ($c = 3.00 \times 10^5$ km/s), the corrected equation is $T - \dfrac{D_1 - D_2}{c} = P + \dfrac{\alpha P}{360}$.

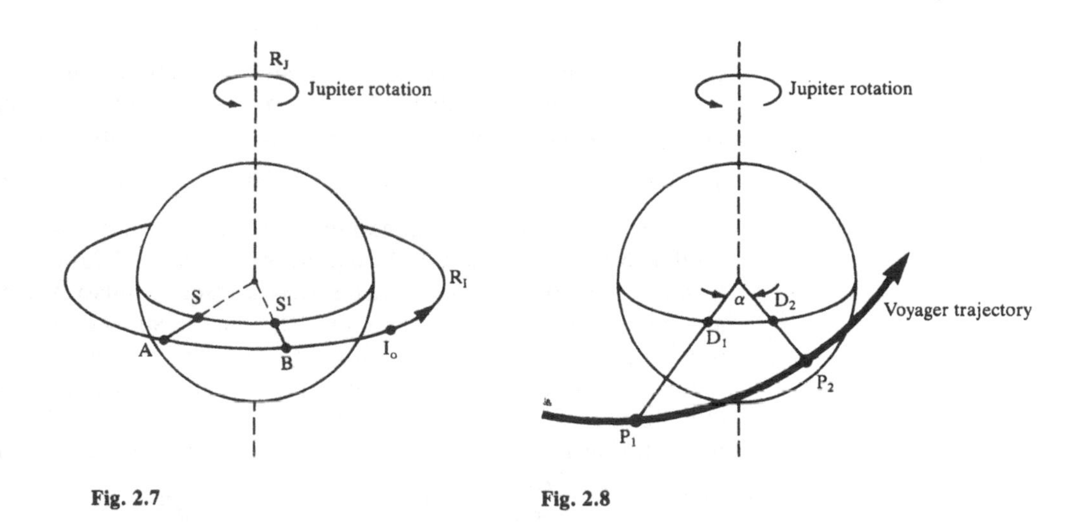

Fig. 2.7 **Fig. 2.8**

For the data provided, we note that

$$\frac{D_1 - D_2}{c} = \frac{(7.70 - 4.47) \times 10^5}{3.00 \times 10^5}\ \text{s}$$

$$\doteq 1 \text{ second.}$$

Since the uncertainty (possible error) is one minute in both time measurements, we see that in this case there is no point in making this correction. The equation $T = P + \frac{\alpha P}{360} = P\left(1 + \frac{\alpha}{360}\right)$ may be solved for P:

$$P = \frac{T}{1 + \frac{\alpha}{360}}.$$

Substituting the data,

$$T = t_2 - t_1$$

$$= \left(16\frac{24}{60} - 2\frac{25}{60}\right)\text{h}$$

$$\doteq (16.40 - 2.42)\text{ h, or } 13.98 \text{ hours.}$$

$$1 + \frac{\alpha}{360} = 1 + \frac{147.2}{360}$$

$$= 1.409, \text{ so}$$

$$P = \frac{13.98}{1.409}\text{ h}$$

$$= 9.92 \text{ hours}$$

c. The observations of the radio emissions from Saturn were much more erratic. A series of measurements $t_1, t_2, \ldots, t_N$ of the times of certain characteristic peak emissions was recorded, but it was not known how many complete rotations had actually occurred between each consecutive pair (t_{i-1}, t_i). The correction $\frac{D_1 - D_2}{c}$ is not negligible in this case, and other corrections not discussed here also apply. Under these circumstances the equation relating T, P, and α becomes impossible to solve directly. However, the iterative approach illustrated next can be used, and by applying it to the case of part (b), we can show that the same solution is obtained. We let $t_{2c} = t_1 + P$. Since P is Jupiter's period, t_{2c} can be considered a "corrected time" in the sense that if *Voyager's* revolution period around Jupiter matched Jupiter's rotation period, we would have $\alpha = 0$, $P = T$, and $t_{2c} = t_2$. Substituting $T = t_2 - t_1$, and $P = t_{2c} - t_1$ for the first term on the right in the equation $T = P + \frac{\alpha P}{360}$, we get $t_2 - t_1 = t_{2c} - t_1 + \frac{\alpha P}{360}$.

Solving for t_{2c}, we get $t_{2c} = t_2 - \frac{\alpha P}{360}$. Since this equation still contains P, the quantity we are seeking, we begin by making an initial guess (P_0) at its value; this is used together with the known values of t_2 and α to find a first estimate of t_{2c}, which we may call t_{2c1}. Next we let $P_1 = t_{2c1} - t_1$ and repeat the evaluation of t_{2c} to get t_{2c2}; then $P_2 = t_{2c2} - t_1$. This process is repeated until the difference $P_n - P_{n-1}$ is less than 1 min $\doteq 0.02$ h, the possible error of our time observation. This procedure is easily done by computer (and in this case may even be done using a hand calculator). Write a computer program to perform this iteration.

Solution: We display a program written in BASIC.

```
]LIST

10  REM  JUPITER PERIOD ITERATION

20  REM  ROUNDED TO 3 DECIMAL PLA
    CES
110 T1 = 2.417
120 T2 = 16.4
130 P(0) = T2 - T1
135 PRINT "M";: HTAB 6: PRINT "T
    C(M)","P(M)
140 FOR M = 1 TO 30
150 TC(M) = T2 - 147.2 * P(M - 1)
    / 360
160 P(M) = TC(M) - T1
164 X% = 1000 * P(M):X(M) = X% /
    1000
166 Y% = 1000 * TC(M):Y(M) = Y% /
    1000
170 PRINT M;: HTAB 6: PRINT Y(M)
    ,X(M)
180 IF ABS (P(M) - P(M - 1)) <
    .017 THEN 200
190 NEXT M
200 PRINT "AFTER ";M;" ITERATION
    S, JUPITER'S PERIOD IS FOUND
     TO BE ";X(M)
210 END
```

```
]RUN
M    TC(M)       P(M)
1    10.682      8.265
2    13.02       10.603
3    12.064      9.647
4    12.455      10.038
5    12.295      9.878
6    12.36       9.943
7    12.334      9.917
8    12.345      9.928
AFTER 8 ITERATIONS, JUPITER'S
PERIOD IS FOUND TO BE 9.928
```

CHAPTER THREE

ALGEBRA

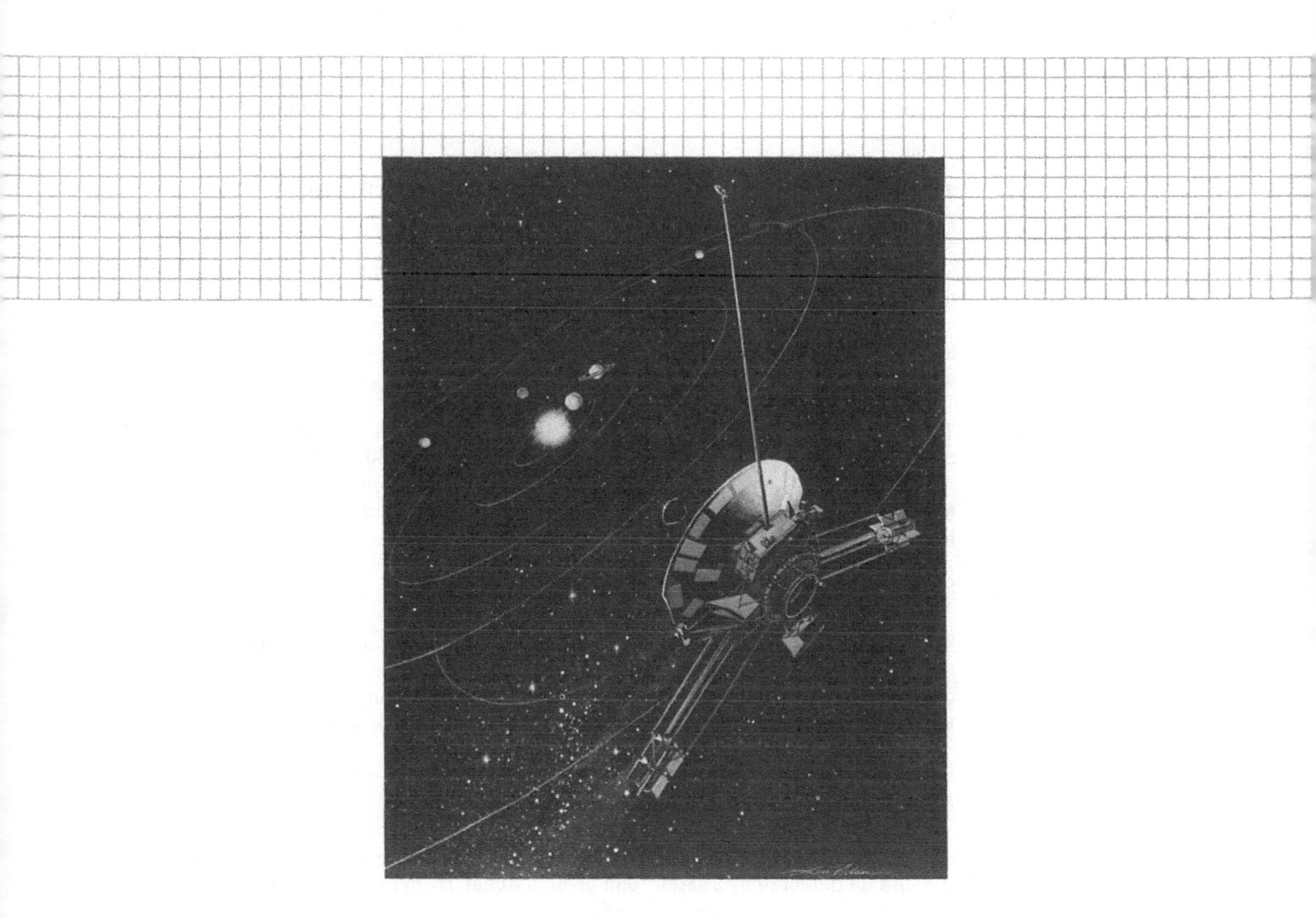

Artist's concept of *Pioneer 10* leaving the solar system, June 13, 1983.

Algebra is the language of quantitative science. As such, its methods and techniques can be found in most of the examples in this volume. The problems selected for this chapter are those that do not also draw heavily on other mathematical areas. Several use the distance, time, and rate relationship, and others use direct and inverse variation. Some approximation techniques that are frequently used to solve otherwise intractable problems are also included.

During 1982, the planets Jupiter and Saturn were in conjunction, so that they appeared very close to each other in the night sky. In the problem that follows, we see how frequently such an event happens.

PROBLEM 1. The planets Earth, Jupiter, Saturn, and Uranus revolve around the Sun approximately once every 1, 12, 30, and 84 years respectively.

a. How often will Jupiter and Saturn appear close to each other in the night sky as seen from Earth?

Solution: The time required must be a multiple of both 12 years and 30 years. This event will recur at intervals of the least common multiple of 12 and 30, or 60 years.

b. How often will Jupiter, Saturn, and Uranus all appear in the same area in the night sky as seen from Earth?

Solution: Now we need the least common multiple of 12, 30, and 84, which is 420 years.

In addition to the electromagnetic radiation that we know as heat and light, the Sun continuously sends out charged particles known as the solar plasma (see also Chapter 10, Problem 1). From time to time, there is a strong burst of highly energetic particles called a solar flare from a small source in the Sun's atmosphere.

PROBLEM 2. When a solar flare occurs on the Sun, it can send out a blast wave that travels through interplanetary space at a speed of 3×10^6 km/h.

a. How long would it take for such a solar flare blast wave to get to Earth where it could be detected by a satellite in orbit? (Recall from the preceding chapter that for a satellite close enough to Earth, we can use the Earth–Sun distance of 1.5×10^8 km for the satellite–Sun distance.)

Solution: Since distance = speed × time,

$$1.5 \times 10^8 \text{ km} = (3 \times 10^6 \text{ km/h})\ (\text{time}).$$

$$\text{Then time} = \frac{1.5 \times 10^8}{3 \times 10^6}\ \text{h} = 0.5 \times 10^2\ \text{h},$$

or 50 h (about two days).

b. When such a solar flare is detected, it is interesting to study the source. Since the Sun is rotating, we must determine how far the source has turned between the emission and the detection of a solar flare. Because the Sun is a dense gas rather than a solid body, it does not have a uniform rotation rate; on the average the Sun makes one complete revolution in 25.4 days. How many degrees would it rotate (on the average) during the time the blast wave traveled to the orbiting satellite?

Solution: Since the Sun rotates 360° in 25.4 days, it rotates $\frac{360}{25.4}$ °/day, so

$$\text{the solar rotation rate} = 14.2\ °/\text{day}$$

$$= \frac{14.2°}{\text{day}} \times \frac{1\ \text{day}}{24\ \text{hour}}$$

$$= 0.59\ °/\text{h}.$$

In 50 hours, the Sun rotates 50 h × 0.59°/h = 29.5°.

PROBLEM 3. A scientific capsule was carried aloft by a rocket and released at the peak of the rocket's trajectory. The rocket had an average vertical speed of 920 km an hour from liftoff to release of the capsule. The capsule made a controlled descent with an average vertical speed of 390 km an hour and landed 68 minutes after the rocket was launched. Find the maximum height reached by the rocket.

Solution: Let t be the time of ascent in hours. Then $\frac{68}{60} - t$ is the time of descent, and since the distances of ascent and descent are equal,

$$920\,t = 390\left(\frac{68}{60} - t\right)$$

$$= 442 - 390\,t$$

$$1310\,t = 442$$

$$t = 0.34\ \text{h}.$$

The maximum height = distance of ascent = (920 km/h) (0.34 h) = 310 km.

Italian scientist Galileo Galilei's introduction of the telescope for studying the heavens brought about a revolutionary change in astronomy. It is expected that a comparable leap in our ability to examine the universe will take place when the Hubble Space Telescope is launched into orbit by the Space Shuttle in 1986. Because the space telescope will be above Earth's atmosphere, it will be able to see much fainter objects than can now be seen by the best Earth-based telescopes. As we shall see in the next problem, this means that astronomers will soon be able to make observations and to compare them with the cosmological theories about the age and formation of the universe.

PROBLEM 4. **a.** The Space Telescope will be able to see stars and galaxies whose brightness is only $\frac{1}{50}$ of the faintest objects now observable using ground-based telescopes. The brightness of a point source such as a star varies inversely as the square of its distance from the observer. How much farther into the universe will the space telescope be able to see compared to ground-based telescopes?

Solution: Let d_G be the distance from Earth of the faintest object visible to a ground-based telescope, and let B_G be the brightness of this object. Let d be the distance of an object of brightness $\frac{1}{50} B_G$. Since brightness varies inversely as the square of distance, $B_G = \frac{k}{d_G^2}$ and $\frac{1}{50} B_G = \frac{k}{d^2}$. Then $\frac{k}{B_G} = d_G^2 = \frac{d^2}{50}$, so $d^2 = 50\, d_G^2$ or $d = 7.1\, d_G$.

The Space Telescope will see about seven times farther.

b. Because of the time it takes for light to travel from distant stars and galaxies, we see them as they were some time ago—the photons that reach us from an object that is 1 parsec away were actually emitted 3.26 years ago (see Chapter 2, Problem 12). The best ground-based telescopes can see objects about 10^9 parsecs from our solar system. How long ago were the photons emitted that we now see when we observe such an object?

Solution: Since 10^9 parsecs $= 3.26 \times 10^9$ light-years, the photons were emitted 3.26×10^9 years ago.

c. When the Space Telescope begins its observations, how far back in time will it see stars and galaxies?

Solution: Since it will see 7.1 times farther, it will see photons that were emitted $7.1 \times 3.26 \times 10^9 = 2.2 \times 10^{10}$ years ago. (If, as suggested by cosmological theory, the age of the universe is between 10 and 20 billion years, the space telescope should enable us to see stars and galaxies in the earliest stages of formation.)

Pioneer 10 was launched on 3 March 1972. It outlived and outperformed the fondest dreams of its creators. Designed to last at least 21 months, it has continued well beyond the accomplishment of its mission. On 25 April 1983, its distance from Earth equaled that of Pluto, and the following June it crossed Neptune's orbit and left the solar system. (Although Pluto is normally the outermost planet in the solar system, it has a highly eccentric orbit and will be closer to the Sun than Neptune will be for the next seventeen years.) To add to its record of endurance, most of *Pioneer 10*'s instruments are still working, and Earth-based tracking stations were still receiving signals bearing information about the behavior of the Sun's extended atmosphere as of this writing.

PROBLEM 5. **a.** How long did *Pioneer 10*'s radio signals, traveling at the speed of light (3.00×10^5 km/s) take to reach Earth from the distance of Pluto in April 1983 (4.58×10^9 km)?

Solution:

$$\text{time} = \frac{\text{distance}}{\text{speed}} = \frac{4.58 \times 10^9}{3.00 \times 10^5}\ \text{s}$$

$$= 1.53 \times 10^4\ \text{s}$$

$$= \frac{1.53 \times 10^4}{3600}\ \text{h, or 4.25 hours}$$

b. What was *Pioneer 10*'s average speed, in km/h, if it traveled about 4.58×10^9 km between 3 March 1972 and 25 April 1983?

Solution: From 3 March 1972 to 3 March 1983 there were 11 years, of which 2 were leap years, and from 3 March 1983 to 25 April there were another 53 days. The time for *Pioneer 10* to travel that distance was $(365 \times 11) + 2 + 53 = 4070$ days, or $4070 \times 24 = 97\,680$ hours.

$$\text{Average speed} = \frac{4.58 \times 10^9}{9.77 \times 10^4}\ \text{km/h} = 4.69 \times 10^4\ \text{km/h}$$

(We note that the average speed over this period is less than the average speed over the 21-month period of Problem 6 in Chapter 2.)

The time required for an orbiting satellite to make one complete revolution of Earth is called its *period*. The length of the period depends on the location of the observer making the measurement.

Suppose the observer is located far out in space and views the satellite against the background of fixed stars. The period measured in this manner is called the *sidereal* period of revolution, or the period in relation to the stars. Note that the rotation of Earth does not affect the sidereal period. Now suppose that the observer is standing on Earth's equator. A satellite is overhead in low Earth orbit moving directly eastward. When the satellite has made one complete transit of its orbit, it will not yet be overhead for the observer because the rotation of Earth will have carried the observer a distance eastward. The satellite must travel an additional distance to again be over the observer's head. The observer measures the period of the satellite as the time elapsing between successive passes directly overhead. This period is called the *synodic* period of revolution, or the period between successive conjunctions, and it takes into account the rotation of Earth.

Spacecraft usually orbit in the same easterly direction as Earth's rotation: this is called a *posigrade* orbit. All U.S. manned spaceflights have been launched in posigrade orbits to take advantage of the extra velocity given to the spacecraft by Earth's rotation. In this case, the synodic period is greater than the sidereal period.

If the direction of orbiting is westerly, or opposite to Earth's rotation, the orbit is said to be *retrograde*. In this case, an Earth observer would meet the satellite before it made one complete revolution around Earth, and the synodic period would be less than the sidereal period.

PROBLEM 6. In Chapter 9, Problem 7, we show that the sidereal period (in seconds) can be computed by the formula $P = 2\pi\sqrt{a^3/GM}$, where a is the average radius of orbit from the center of the body about which the satellite is in motion, G is the constant of universal gravitation, and M is the mass of the body about which the satellite orbits.

a. Find the sidereal period of the *High Energy Astronomy Observatory* (HEAO) satellites, which have an average altitude above Earth of 430 km. The radius of the Earth averages 6370 km, and the value of the product GM for Earth is $3.99 \times 10^{14}\ \text{m}^3/\text{sec}^2$.

Solution: The radius of orbit is the sum of the radius of Earth and the average altitude of the satellite:

$$a = 6370\ \text{km} + 430\ \text{km} = 6.80 \times 10^6\ \text{m}.$$

Then the sidereal period in seconds is

$$P = 2(3.142)\sqrt{\frac{(6.80 \times 10^6)^3}{3.99 \times 10^{14}}}$$

$$P = (6.284)\,(6.80)\sqrt{\frac{6.80}{3.99}} \times 10^2 = 5580\ \text{s}$$

The sidereal period then is 93.0 minutes, or 1.55 hours.

b. Compute the synodic period of the HEAO satellites, given that their orbits are posigrade.

Solution: In Fig. 3.1, let x be the position of the observer (assumed on the equator) when the satellite is directly overhead and let y be the observer's position one synodic period later, due to the rotation of Earth.

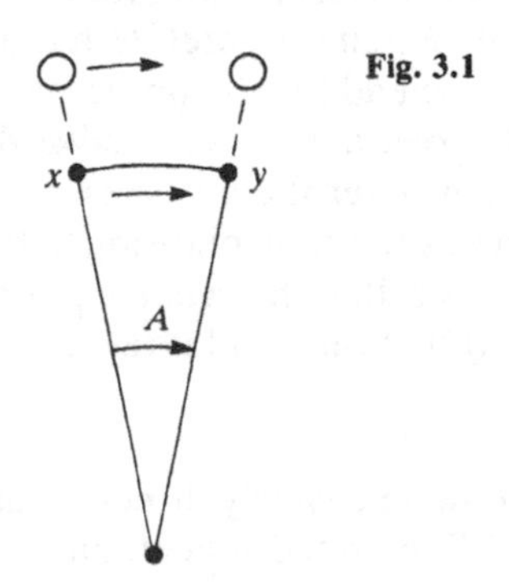

Fig. 3.1

If we can find the angular distance A, we shall be able to use it to find the synodic period. In one synodic period the observer traveled an angular distance A, and the satellite traveled an angular distance $360° + A$, measuring the angular distance in degrees. The observer travels 360° in 24 hours, or 1° in 24/360 hours, so it takes the observer $(24/360)(A)$ hours to travel the angular distance A. From part (a), the sidereal period is 1.55 hours. It takes the satellite 1.55/360 hours to travel 1°, and the time that elapses between successive viewings over the observer is therefore $\left(\frac{1.55}{360}\right)(360 + A)$ hours.

$$\frac{24}{360} A = \frac{1.55}{360} (360 + A)$$

$$24 A = (1.55)(360) + 1.55 A$$

$$22.45 A = 558$$

$$A = 24.9 \text{ degrees}$$

So the synodic period is $\left(\frac{1.55}{360}\right)(360 + 24.9) = 1.66$ hours $= 99.6$ minutes. We observe that the synodic period is 6.6 minutes longer than the sidereal period.

PROBLEM 7. The statement has been made that Newton's derivation of his inverse-square law of gravity from Kepler's third law is among the most important calculations ever performed in the history of science. Kepler's third law, based on observation rather than theory, states that the squares of the periods of any two planets are to each other as the cubes of their average distances from the Sun. Derive Newton's law from Kepler's law.

Solution: If we represent the periods of any two planets by t and T and their distances from the Sun by r and R, respectively, then

$$\frac{T^2}{t^2} = \frac{R^3}{r^3},$$

or

$$T^2 = \frac{t^2 R^3}{r^3}.$$

Assuming that we know the values of t and r, and substituting a constant C for the quantity $\frac{t^2}{r^3}$ the equation can be reduced to

$$T^2 = CR^3.$$

Thus if we know either T or R for the second planet, we can solve for the unknown quantity. In this problem, however, we wish to use this equation to discover a new relationship, Newton's law of gravitation. For a body moving in a circular path, the acceleration toward the center is

$$a = \frac{v^2}{r}.$$

Substituting in $F = ma$,

$$F = \frac{mv^2}{r}.$$

The velocity of the body in the circular orbit is

$$v = \frac{2\pi r}{T}.$$

So

$$F = \frac{mv^2}{r} = \frac{m}{R}\left(\frac{2\pi R}{T}\right)^2$$

Because $T^2 = CR^3$, we find by substitution in the previous equation that

$$F = \frac{4\pi^2 m}{C} \times \frac{1}{R^2}$$

$$= \frac{K}{R^2}.$$

That is, the force holding a planet in orbit falls off as the square of the distance R to the Sun. Newton expressed the value of K and obtained his law of universal gravitation:

$$F = \frac{GMm}{r^2}.$$

This law applies not only to the attraction between a planet and the Sun but also to the attraction between any two bodies. G is the constant of universal gravitation, M and m are the masses of the two bodies, and r is the distance between their centers of mass.

In solving the next problem, two special techniques are needed. One is a frequently used approximation based on the fact that $(1 + x)(1 - x) = 1 - x^2$. If x is small (for example, suppose $x = 0.01$), then x^2 is very much smaller (for $x = 0.01$, $x^2 = 0.0001$), and in this case it is well within the limits of experimental error to use $(1 + x)(1 - x) = 1$, or $\frac{1}{1 + x} = 1 - x$. The other is the substitution of a single variable for the ratio of two other variable names.

PROBLEM 8. From Kepler's third law, we see that the farther a planet is located from the Sun, the longer its period is. Suppose Earth orbits the Sun in a circle of radius r ($r = 1.5 \times 10^8$ km) with a period T ($T = 1$ year). Then any spacecraft SC (see Fig. 3.2) orbiting the Sun in the same plane but at some greater distance $(r + a)$ will have a period larger than T, and if it starts from a point on the extension of the Earth–Sun line (as shown), it will gradually lag farther and farther behind.

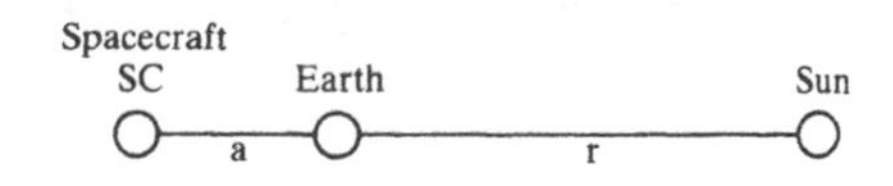

Fig. 3.2

However, the situation changes if a is sufficiently small, because then the gravity of Earth, in the configuration shown, adds appreciably to that of the Sun. For the force holding the spacecraft in orbit to balance the *combined* pull of the Earth and Sun, the spacecraft must move a bit faster. In fact, there is a particular value for a so that the speeding up of the spacecraft is just sufficient to allow it to keep up with Earth. If that happens, then the spacecraft orbits the Sun in a circle of radius $(r + a)$, but with period T like Earth. What is the value of a that allows such an orbit?

Solution: Let m_s, m_e, m_{sc} be the masses of the Sun, Earth, and spacecraft, respectively. For Earth's motion, we have, as in the foregoing problem,

$$F = G\left(\frac{m_s m_e}{r^2}\right) = \frac{m_e v_e^2}{r} = \frac{m_e}{r}\left(\frac{2\pi r}{T}\right)^2$$

or

$$\frac{Gm_s}{r^3} = \left(\frac{2\pi}{T}\right)^2 \tag{1}$$

For the motion of the spacecraft, similar analysis gives

$$F = G\left(\frac{m_{sc}m_e}{a^2} + \frac{m_{sc}m_s}{(r+a)^2}\right) = \frac{m_{sc}v_{sc}^2}{r+a} = \frac{m_{sc}}{r+a}\left(\frac{2\pi(r+a)}{T}\right)^2 = m_{sc}(r+a)\left(\frac{2\pi}{T}\right)^2.$$

Canceling m_{sc} and substituting $\frac{Gm_s}{r^3} = \left(\frac{2\pi}{T}\right)^2$ from (1):

$$G\left(\frac{m_e}{a^2} + \frac{m_s}{(r+a)^2}\right) = (r+a)\left(\frac{Gm_s}{r^3}\right)$$

and dividing by $G \cdot m_s$:

$$\frac{m_e}{m_s}\frac{1}{a^2} + \frac{1}{(r+a)^2} = \frac{r+a}{r^3} = \frac{1+(a/r)}{r^2}.$$

Now let $u = \frac{m_e}{m_s}$ and $z = \frac{a}{r}$. Notice that z is very small, since a is much less than r. After these substitutions,

$$\frac{u}{a^2} + \frac{1}{(r + a)^2} = \frac{1 + z}{r^2}.$$

If we divide each denominator by r^2 and again use $z = \frac{a}{r}$,

$$\frac{u}{z^2} + \frac{1}{(1 + z)^2} = 1 + z.$$

Although this equation contains only u and z, it still has no simple solution, so we now make two approximations in the second term on the left:

$$\frac{1}{(1 + z)^2} \doteq (1 - z)^2 = 1 - 2z + z^2 \doteq 1 - 2z:$$

$$\frac{u}{z^2} + (1 - 2z) = 1 + z, \text{ simplifying to } \frac{u}{z^2} = 3z, \text{ or } z^3 = \frac{u}{3}$$

Now the quantity $u = \frac{m_e}{m_s} = 3 \times 10^{-6}$, so $z^3 = 10^{-6}$, or $z = 10^{-2}$.

Since $z = \frac{a}{r}$, $a = r \times 10^{-2} = 1.5 \times 10^6$ km.

The position we have found in this problem is an equilibrium point of the Sun-Earth system. A similar analysis can be used to show the existence of another equilibrium point on the sunward side of Earth, and in fact there are five such equilibrium points in all for any two-body gravitational system. These are called Lagrangian points in honor of the mathematician who first proposed their existence (Fig. 3.3). It has been suggested that two of the Lagrangian points of the Earth-Moon system should be considered as possible locations for future space colonies.

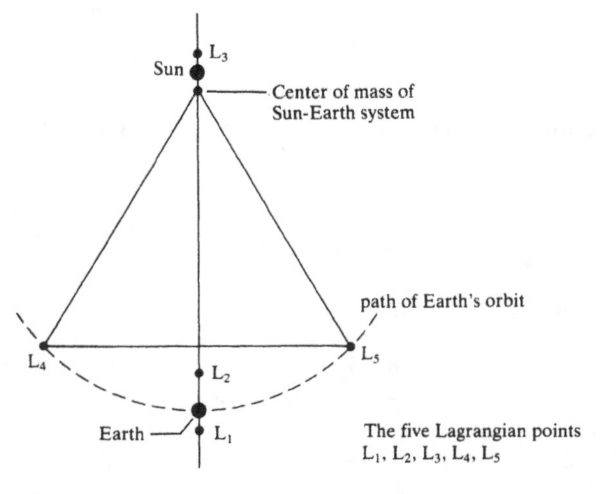

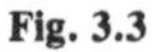
Fig. 3.3

The ISEE-3 satellite (third *International Sun-Earth Explorer*), a joint venture of NASA and the European Space Agency, was launched in August 1978 and placed in a "halo orbit" around the Lagrangian equilibrium point between Earth and the Sun. In this orbit, it monitored the Sun's emissions that approach Earth—without the interference that would result if the satellite were actually at the equilibrium point where its radio antenna would have to point directly at the Sun. After this mission was successfully completed in 1982, ISEE-3's orbit and direction were changed to conduct an exploratory survey of Earth's magnetotail. In December 1983, the satellite was redirected toward the comet Giacobini-Zinner and renamed *International Cometary Explorer* (ICE) in keeping with its new mission. It reached this comet in September 1985.

PROBLEM 9. **a.** If M is the mass of Earth, then the mass of the Moon is $0.012M$. The radii of Earth and the Moon are 6370 and 1740 km, respectively. Use these facts with Newton's law of universal gravitation to find the ratio of surface gravity on the Moon to surface gravity on Earth.

Solution: If we place a mass m at the surface of Earth, then the gravitational attraction between the mass and Earth is

$$F_e = \frac{GMm}{(6370)^2}.$$

Similarly, the attraction between the Moon and an equal mass m placed on its surface is

$$F_m = \frac{G(0.012M)m}{(1740)^2}.$$

The ratio of F_m to F_e is

$$\frac{F_m}{F_e} = \frac{0.012}{(1740)^2} \times \frac{(6370)^2}{1}$$

$$= \frac{4.87 \times 10^5}{3.03 \times 10^6}$$

$$\doteq \frac{1}{6}.$$

That is, gravity at the surface of the Moon is $\frac{1}{6}$ as great as gravity at the surface of Earth.

b. If a man weighs 180 pounds on Earth, what would he weigh on the Moon?

Solution: Weight on the Moon would be as follows:

$$\frac{1}{6} \times 180 \text{ lb} = 30 \text{ lb}$$

PROBLEM 10. In some future space stations it is expected that artificial gravity will be created by rotation of all or part of the station. Gas jets or other propulsion devices can be used to control the rate of rotation of the station. As with the centrifuge, the rotation will produce a force against the astronaut that cannot be distinguished from gravity. If r is the distance of a point in the station from the center of rotation, then the velocity of the point for N rotations a second is

$$v = 2\pi rN.$$

As noted in Problem 7,

$$a = \frac{v^2}{r}, \text{ or } v = \sqrt{ar}.$$

Setting the two velocities equal,

$$2\pi rN = \sqrt{ar}$$

$$N^2 = \frac{ar}{(2\pi)^2 r^2}$$

$$N = \frac{1}{2\pi}\sqrt{\frac{a}{r}}.$$

If r is given in meters, then a is the acceleration in meters per second per second. By controlling the values of r and N, any desired artificial gravity can be produced.

a. Compute the rotational rate needed if the radius of the station is 30 m and a gravity equal to one-half the gravity of Earth is desired. (Use g = 980 cm per second per second, or 9.8 m per second per second.)

Solution:

$$a = \frac{1}{2}(9.8 \text{ m/s}^2) = 4.7$$

$$N = \frac{1}{2\pi}\sqrt{\frac{4.7}{30}}$$

$$= \frac{\sqrt{0.157}}{2\pi}$$

$$= 0.063$$

The rate of rotation must be 0.063 rotation per second or 60 × 0.063 = 3.8 rotations per minute.

b. Compute the needed rotational rate if the radius of the station is 150 m and Earth surface gravity is desired.

Solution:

$$N = \frac{1}{2\pi}\sqrt{\frac{9.8}{150}}$$

$$= \frac{\sqrt{.065}}{2\pi}$$

$$= 0.04$$

The rate of rotation must be 0.04 rotation per second or 2.4 rotations per minute.

PROBLEM 11. **a.** The force of gravitation with which one body attracts another is inversely proportional to the square of the distance between them. Consequently, the pull of the Moon on the oceans is greater on one side of Earth than on the other. This gravitational imbalance produces tides. The Sun affects the tides similarly. Because the Sun exerts an enormously greater pull on Earth than the Moon does, one might think that the Sun would influence the tides more than the Moon. Just the opposite is true. How can this be?

Solution: Let N be the point on Earth nearest the Moon and let F be the point on Earth farthest from the Moon. We shall assume that the tide-raising force of the Moon is in some sense measured by the difference in the pull of the Moon on unit masses located at N and F (see Fig. 3.4). If r is the distance from the center of the Moon to N and if D_e is the diameter of Earth, then the forces acting at N and F are, respectively, $\frac{GM}{r^2}$ and $\frac{GM}{(r + D_e)^2}$, M being the mass of the Moon and G the universal gravitational constant. The difference between these two forces is the tide-raising force, which we shall call F_t. Then,

$$F_t = GM\left(\frac{1}{r^2} - \frac{1}{(r + D_e)^2}\right)$$

$$= \frac{2GMD_e\left(1 + \frac{D_e}{2r}\right)}{r^3\left(1 + \frac{D_e}{r}\right)^2}.$$

Because $\frac{D_e}{r}$ is very small, this expression is approximately

$$F_t = \frac{2GMD_e}{r^3}.$$

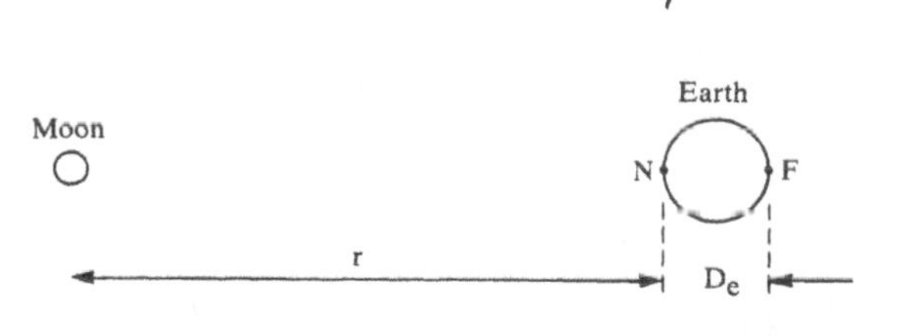

Fig. 3.4

Thus we would expect the tidal effect to be inversely proportional to the cube of the distance, whereas gravity is inversely proportional to the square of the distance. Because the distance from Earth to the Sun is enormously greater than the distance to the Moon, it is not surprising that the Moon provides the dominant tide-raising force. Local horizontal components of this force cause the tides to roll in and roll out (i.e., the horizontal movement of the water).

b. From the foregoing, we can compare the tide-raising forces of the Moon and the Sun. If we use the subscript *m* for variables that apply to the Moon and *s* for those that apply to the Sun, the ratio

$$\frac{F_{tm}}{F_{ts}} = \frac{2GM_mD_e/r_m^3}{2GM_sD_e/r_s^3} = \frac{M_mr_s^3}{M_sr_m^3}.$$

The mass and distance of the Moon and Sun are as follows:

$$M_m = 73.5 \times 10^{21}\text{kg};\ M_s = 1.99 \times 10^{30}\text{kg}$$

$$r_m = 3.84 \times 10^5\text{km};\quad r_s = 1.5 \times 10^8\text{km}$$

Compute $\frac{F_{tm}}{F_{ts}}$.

Solution:

$$\frac{F_{tm}}{F_{ts}} = \frac{73.5 \times 10^{21} \times (1.5 \times 10^8)^3}{1.99 \times 10^{30} \times (3.84 \times 10^5)^3} = \frac{248}{113} \times 10^{45-45} \doteq 2.2$$

So the tidal force exerted by the Moon is more than double that exerted by the Sun on the Earth.

CHAPTER FOUR

GEOMETRY

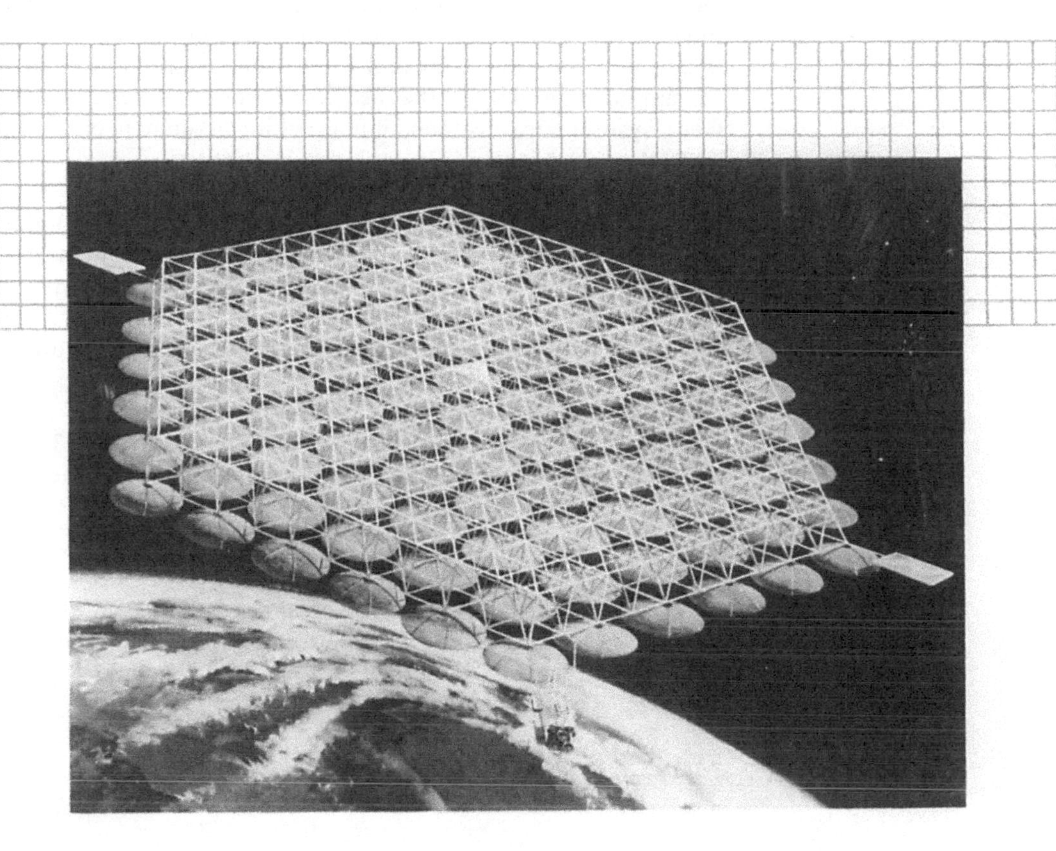

A hypothetical space station with "electronic mail" capabilities.

Geometry is fundamental to space science. A multitude of activities, from the prediction of flight paths to the design of equipment, make use of geometric analysis. Geometry enters into many of the problems of the preceding and subsequent chapters. Most of the problems in this chapter fall into three categories: those involving areas and volumes of plane and solid figures, those that use similarity, and those that use properties of circles or spheres. The Sun-Earth-Moon system happens to exhibit a striking geometric coincidence, which we examine in the first problem.

PROBLEM 1. To an observer on Earth, the Sun and the Moon subtend almost the same angle in the sky. The average angle is 0.52 degrees for the Moon and 0.53 degrees for the Sun. Depending on the particular location in its elliptic orbit, the Moon's angle ranges between 0.49° and 0.55°, whereas that of the Sun ranges between 0.52° and 0.54°. This is why the Moon sometimes completely blocks the Sun, producing a total solar eclipse.

a. If the mean lunar and solar distances are respectively 3.8×10^5 km and 1.5×10^8 km, what is the ratio of the solar diameter to the lunar diameter, and what is the ratio of the solar volume to the lunar volume?

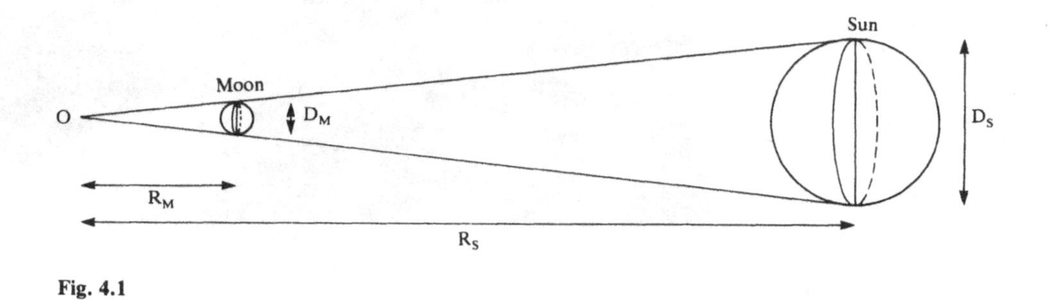

Fig. 4.1

Solution: The geometry of the eclipse is illustrated in Fig. 4.1. Since the angle at 0 is the same for both the large and the small triangles and the triangles are isosceles, they must be similar. Letting R_M and R_S denote the lunar and solar distances, respectively, and D_M and D_S the lunar and solar diameters, we have

$$\frac{D_S}{D_M} = \frac{R_S}{R_M} = \frac{1.5 \times 10^8}{3.8 \times 10^5} = 390.$$

If V_M and V_S are the lunar and solar volumes, respectively,

$$\frac{V_S}{V_M} = \frac{(4/3)\pi(D_S/2)^3}{(4/3)\pi(D_M/2)^3} = \left(\frac{D_S}{D_M}\right)^3 = (390)^3 = 5.9 \times 10^7.$$

b. Since the angle at 0 is very small, we can approximate the lunar or solar diameter by the arc length of the circle with radius R_M or R_S where this arc length subtends the angle at 0. Use the relation $s = r\theta$ (θ in radians) to determine the actual values of D_M and D_S.

Solution: $0.52° = 0.52° \times \frac{\pi \text{ rad}}{180°} = 0.0091 \text{ rad}$

$$D_M = R_M(0.0091) = 3.8 \times 10^5 \times 0.0091 = 3.5 \times 10^3 \text{ km}$$

$0.53° = 0.0092$ rad, by the same conversion just shown

$$D_S = R_S(0.0092) = 1.5 \times 10^8 \times 0.0092 = 1.4 \times 10^6 \text{ km}$$

(*Note:* The reader may prefer to avoid the approximation by using the tangent function—that is, $\tan\frac{\theta}{2} = \frac{\text{lunar radius}}{\text{lunar distance}} = \frac{\text{solar radius}}{\text{solar distance}}$; note, however, that to two significant digits the result is the same.)

PROBLEM 2. All the energy to meet needs on Earth, whether the energy is natural or synthetic, ultimately comes or has come from the Sun in the form of electromagnetic radiation. There has been much interest recently in using this radiant source of energy directly to supplement or supplant the existing power sources. Further, since our Sun is but one of many stars, it is of interest to compare its energy output with that of other celestial objects.

One measure of the total energy radiated by the Sun received at a unit area of the Earth's surface is called the *solar constant* (where radiation is summed over all wavelengths of the electromagnetic spectrum).

A radiometer flown on the Solar Maximum Mission (SMM) is able to measure accurately the intensity of solar radiation. SMM is a satellite in orbit around Earth at low altitude, and its measurements can be used to provide a good estimate of the solar constant.

The radiometer on SMM admits solar radiation through a small aperture whose area is 0.50 cm^2, and it measures the rate of entrance of this radiation accurately. The spacecraft attitude (pointing direction) is controlled so that the entrance aperture is perpendicular to the line of sight between SMM and the Sun.

a. Over one observation period, radiation entered the radiometer at the rate of 0.069 watts. What is the value of the solar constant, S, as determined by this observation? (Use an extra significant digit in the answer, since this quantity will be used in subsequent calculations.)

Solution:

$$S = \frac{0.069 \text{ watts}}{0.50 \text{ cm}^2} = 0.138 \text{ watts/cm}^2$$

b. It is generally assumed that the Sun emits radiation uniformly in all directions. If this is true, calculate the total rate of energy radiation by the Sun.

Solution: Since the radiation energy rate measurement contains only two significant digits, we can use the Earth-Sun distance of 1.5×10^8 km as SMM's distance from the Sun (see Chapter 2, Problem 8). If the Sun emits uniformly in all directions, the total rate of energy radiation from the Sun is the product of the solar constant and the area of the sphere with radius 1.5×10^8 km, or 1.5×10^{13} cm.

Letting P = total rate of energy radiation from the Sun,

$$P = (S)(4\pi r^2)$$
$$= (0.138 \text{ watts/cm}^2)(4\pi)(1.5 \times 10^{13} \text{ cm})^2$$
$$= 3.9 \times 10^{26} \text{ watts.}$$

c. The foregoing are typical values. Variations of approximately 0.05 percent have been observed at other times. How much do such variations affect S and P?

Solution:

$$\Delta S = 0.05 \times 10^{-2} \times 0.138 = 6.9 \times 10^{-5} \text{ watts/cm}^2$$

$$\Delta P = 0.05 \times 10^{-2} \times 3.9 \times 10^{26} = 2.0 \times 10^{23} \text{ watts}$$

(*Note:* These variations occur on a short time scale (day to day) and are thought to average to zero over a long time scale. A 0.05-percent systematic variation in solar radiation over a time scale of years could produce significant climate changes on Earth.)

d. In 1981, SMM lost pointing accuracy because of a component failure on the spacecraft. Suppose that the orientation of the spacecraft changed so that the line perpendicular (the normal) to the entrance aperture made an angle of 30° with respect to the Sun-SMM line, rather than being parallel to it. By how much would the radiation entering the radiometer be changed?

Solution: For simplicity, let us assume the aperture is a square, ABCD (see Fig. 4.2), with side length a, where $a^2 = 0.50$ cm^2. Looking at this square edge-on with AD as the tilted edge, if DE is parallel to the direction of solar radiation incidence and AE is perpendicular to DE, the aperture is effectively a rectangle whose dimensions are AB and AE. We label the angles $\alpha, \beta, \gamma, \delta$ as shown. Since ($\angle\alpha, \angle\beta$) and ($\angle\gamma, \angle\delta$) are complementary pairs of angles, and since $\angle\beta = \angle\gamma$, we have $\angle\delta = \angle\alpha = 30°$, so ΔADE is similar to the standard 30°-60°-90° triangle, and the ratios $\frac{AD}{2}, \frac{DE}{1}$, and $\frac{AE}{\sqrt{3}}$ are equal, giving AE $= \frac{\sqrt{3}}{2}$AD $= 0.866\, a$. The area of the effective aperture is therefore $(0.866\, a)(a) = 0.866\, a^2$. In other words, the radiometer will register only 87 percent of what it did before losing pointing control.

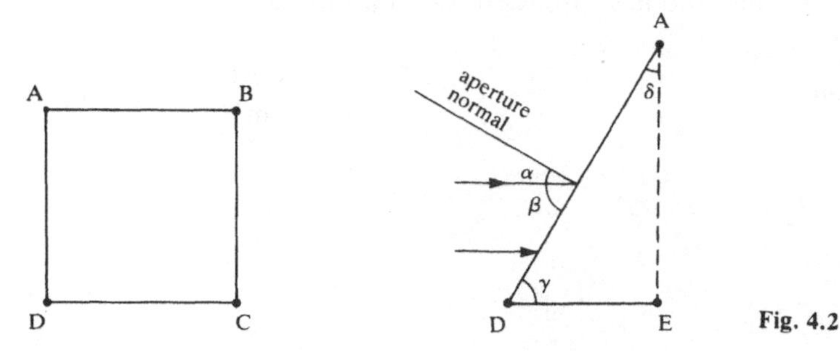

Fig. 4.2

(We observe that the result holds for apertures that are not square.)

As observed in part (c) of the foregoing problem, one of the interesting outcomes of modern advances in the precision with which it is now possible to make measurements of the solar constant is that this quantity is in fact not really a constant!

PROBLEM 3. Solar cells convert the energy of sunlight directly into electrical energy. For each square centimeter of solar cell in direct overhead sunlight, about 0.01 watt of electrical power is available. A solar cell in the shape of a regular hexagon is required to deliver 15 watts. Find the minimum length of a side.

Solution: The total area required is 15 watts/0.01 watt per square centimeter, or 1500 square centimeters. The regular hexagon can be partitioned into six congruent equilateral triangles, each with an area of 1500/6 = 250 square centimeters (see Fig. 4.3).

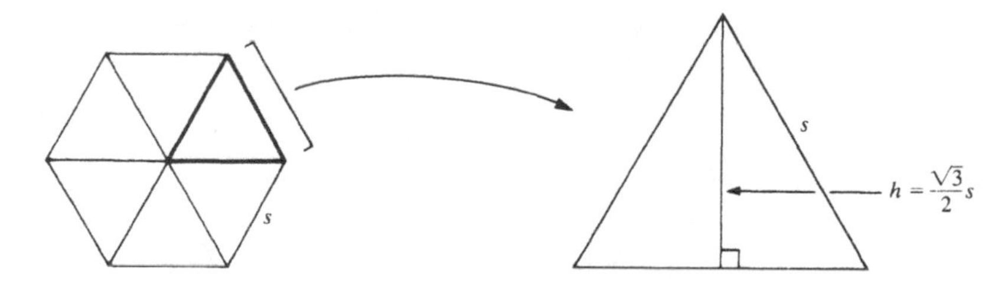

Fig. 4.3

The area A of any equilateral triangle with side s may be expressed $A = \frac{1}{2}(\text{base})(\text{altitude}) = \frac{s}{2} \cdot \frac{\sqrt{3}s}{2} = \frac{\sqrt{3}s^2}{4}$. Solving for s, we have

$$s = \sqrt{\frac{4A}{\sqrt{3}}} = \sqrt{\frac{4(250)\text{ cm}^2}{1.73}} = \sqrt{578\text{ cm}^2} = 24\text{ cm}.$$

PROBLEM 4. Solar cells are made in various shapes to use most of the lateral area of satellites. A certain circular solar cell with radius r will produce 5 watts. Two equivalent solar cells are made, one being a square with side s and the other an equilateral triangle with side p. Find r in terms of p and also in terms of s.

Solution: For the solar cells to have equivalent outputs, their areas must be equal. Thus for the circle and square, we have

$$A_{\text{circle}} = A_{\text{square}}: \pi r^2 = s^2$$

$$r = \frac{s}{\sqrt{\pi}}$$

$$= 0.564\,s.$$

For the circle and equilateral triangle, we have

$$r = p\sqrt{\frac{\sqrt{3}}{4\pi}}$$

$$= 0.371\,p.$$

PROBLEM 5. The largest component of the prelaunch Space Shuttle configuration is the external tank, which serves as the "gas tank" for the Orbiter—it contains the propellants used by the main engines. Approximately 8.5 minutes after launch, when most of the propellants have been used, the external tank is jettisoned. It is the only major part of the Space Shuttle that is not reused.

Fig 4.4(a) shows the launch configuration with the back view of the Orbiter, and Fig. 4.4(b) shows the side view. Fig. 4.5(a), (b), and (c) show the liquid hydrogen tank, the intertank, and the liquid oxygen tank, respectively. The intertank serves as a mechanical connector between the liquid oxygen and liquid hydrogen tanks, and contains the upper dome of the liquid hydrogen tank and the lower dome of the liquid oxygen tank.

a. Using the dimensions provided in the diagrams, estimate the volume of each of the tanks by dividing the tanks into shapes whose volumes are easy to compute.

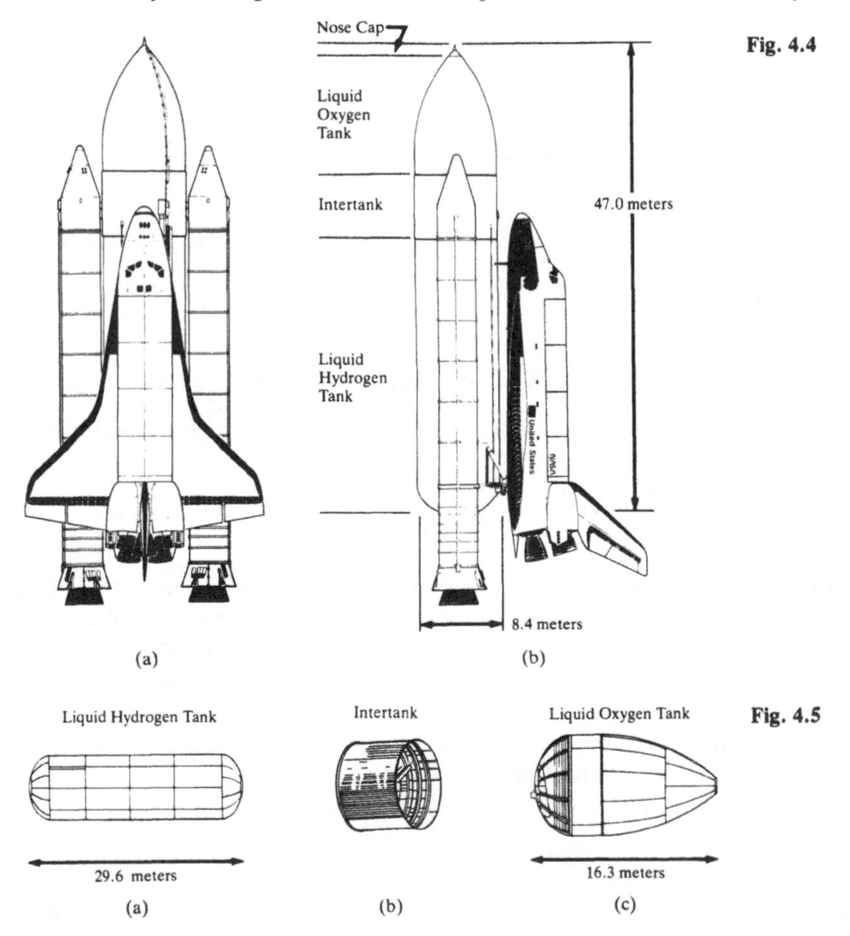

Fig. 4.4

Fig. 4.5

Solution: (This is one possible solution.) The liquid hydrogen tank has the shape of a cylinder with ellipsoidal caps on each end. Since the formula for the volume of a hemisphere is better known, let us approximate the domes as hemispheres. Now the total length of the tank is given as 29.6 m, and the diameter as 8.4 m; our approximation, then, consists of two hemispheres (or a single sphere) of radius 4.2 m and a cylinder of radius 4.2 m and length 29.6 – 8.4 = 21.2 m. The resulting volume estimate is

$$\frac{4}{3}\pi r^3 + \pi r^2 h = \frac{4}{3}\pi (4.2)^3 + \pi (4.2)^2 (21.2)$$

$$= 310 + 1170, \text{ or } 1480 \text{ m}^3.$$

The liquid oxygen tank can be approximated by joining a hemisphere of radius 4.2 m, a cylinder of radius 4.2 m and length about 4 m, and a cone with base radius 4.2 m and height 8.1 m. (This should probably underestimate the volume, since the tapered section is larger than a cone.) Using this dissection, we find that the volume estimate is

$$= \frac{2}{3}\pi r^3 + \pi r^2 h_{\text{cyl}} + \frac{1}{3}\pi r^2 h_{\text{cone}}$$

$$= \frac{2}{3}\pi (4.2)^3 + \pi (4.2)^2 (4) + \frac{1}{3}\pi (4.2)^2 (8.1)$$

$$= 155 + 222 + 150$$

$$= 527 \text{ m}^3.$$

b. The actual volumes of the hydrogen and oxygen tanks, respectively, to the nearest m^3 are 1450 m^3 and 541 m^3. What are the absolute and relative errors of the estimates in (a)? (See Chapter 2 for a discussion of these errors.)

Solution: For the hydrogen tank:

$$\text{Absolute error} = |\text{estimate} - \text{true value}| = |1480 - 1450| = 30 \text{ m}^3$$

$$\text{Relative error} = \frac{\text{absolute error}}{\text{true value}} \times 100\% = \frac{30}{1450} \times 100\% = 2.1\%$$

For the oxygen tank:

$$\text{Absolute error} = |527 - 541| = 14 \text{ m}^3$$

$$\text{Relative error} = \frac{14}{541} \times 100\% = 2.6\%$$

c. The outside of the external tank is covered with a multilayered thermal protective coating to withstand the extreme temperature variations expected during prelaunch, launch, and early flight. Although there are variations in the exact type of material and the thickness at various locations on the tank, the average thickness is 2.5 cm. Estimate the total volume of the insulation material on the tank, assuming a uniform thickness of 2.5 cm.

Solution: A simple way to get such an estimate is to model the external tank as three sections: the lowest section is approximately a hemisphere of radius 4.2 m; the middle section is an open cylinder of radius 4.2 m and height $(47.0 - 4.2 - 8.1) = 34.7$ m; the top part is approximately a cone of base radius 4.2 m and height 8.1 m. The volume of insulation is then close to the product of the surface area of this figure and the thickness 2.5 cm, or 0.025 m.

The surface area of an open hemisphere of radius r is $2\pi r^2$; the lateral area of a cylinder of radius r and length h is $2\pi rh$; the lateral area of a cone of radius r and slant height s is πrs—in this case we know the vertical height h rather than the slant height, but s, h, and r are related by $s^2 = r^2 + h^2$, or $s = \sqrt{r^2 + h^2}$ (see Fig. 4.6).

The surface area, then, is

$$2\pi(4.2)^2 + 2\pi(4.2)(34.7) + \pi(4.2)^2\sqrt{(4.2)^2 + (8.1)^2}$$
$$= 111 + 916 + 120$$
$$= 1147 \text{ m}^2.$$

The volume, then, is $(1147)(.025) = 29 \text{ m}^3$.

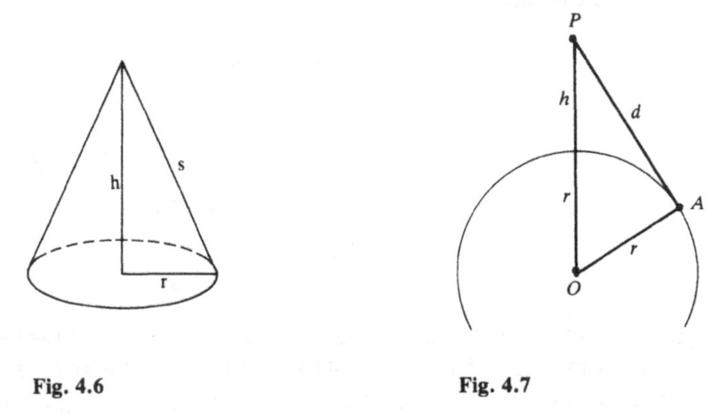

Fig. 4.6

Fig. 4.7

PROBLEM 6. A spacecraft is at P, at an altitude h above Earth's surface, as pictured in Fig. 4.7. The distance to the horizon is d, and r is the radius of Earth.

a. Derive an equation for d in terms of r and h.

Solution: Because PA is tangent to the circle at A, angle PAO is a right angle. Then

$$r^2 + d^2 = (r + h)^2$$
$$d^2 = (r + h)^2 - r^2$$
$$= 2rh + h^2$$
$$d = \sqrt{2rh + h^2}.$$

b. The satellite *Atmospheric Explorer 3* (AE-3) has an elliptic orbit with apogee height 4300 km and perigee height 150 km. Find the distance from AE-3 to the horizon at apogee and perigee. The radius of Earth, to two significant figures, is 6400 km.

Solution: At apogee:

$$d = \sqrt{2(6400)(4300) + (4300)^2}$$
$$= 100\sqrt{5504 + 1849}$$
$$= 8600 \text{ km}$$

At perigee:

$$d = \sqrt{2(6400)(150) + (150)^2}$$
$$= 10\sqrt{19200 + 225}$$
$$= 1400 \text{ km}$$

c. If h is small compared to r, the formula for d found in part (a) can be simplified by dropping the h^2 term, leaving $d = \sqrt{2rh}$. Redo the calculations of part (a) using the simplified formula and compute the relative errors of these approximations.

Solution: At apogee:

$$d_{\text{approx.}} = \sqrt{2(6400)(4300)} = 7400 \text{ km}$$

$$\text{relative error} = \frac{8600 - 7400}{8600} = 0.14 = 14\%$$

At perigee:

$$d_{\text{approx.}} = \sqrt{2(6400)(150)} = 1400 \text{ km}$$

This agrees with the previous result, and the relative error is 0.

d. For what range of values of h is the approximation $d = \sqrt{2rh}$ accurate to two significant digits?

Solution: We need to know the range of values of h that satisfy the following condition:

$$\sqrt{2rh + h^2} < \sqrt{2rh} + 0.01\sqrt{2rh}$$

$$\sqrt{2rh + h^2} < 1.01\sqrt{2rh}$$

$$2rh + h^2 < (1.01)^2(2rh) = 2rh\,(1.0201)$$

$$h^2 < 0.04rh$$

$$h < 0.04r$$

For $r = 6400$ km, we need $h < 256$ km.

From the foregoing, we see that under certain conditions it is possible to substitute a simple formula for a complicated one without affecting the results. Great care must be taken, of course, to ensure that the conditions needed for such simplification are in fact satisfied. Another useful result based on two such approximations is developed in Problem 8. But first we consider the basic geometry of photographic scale.

PROBLEM 7. In Fig. 4.8, the flight path of an airplane or satellite carrying a camera with its lens at C is shown by the arrow. The camera is at a height H above the ground and has focal length f.

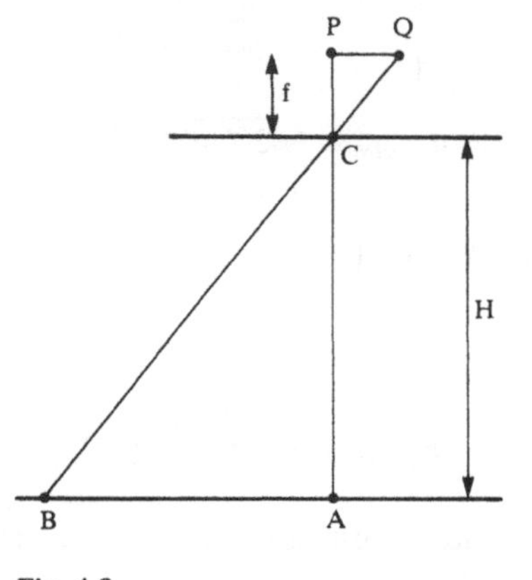

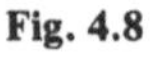
Fig. 4.8

a. If PQ is the image on the film of line AB on the ground, find the scale of the picture, the ratio $\frac{PQ}{AB}$, assuming the picture is taken vertically (PA is perpendicular to both the film and the ground).

Solution: Since triangles ABC and PQC are similar, $\frac{PQ}{AB} = \frac{f}{H}$.

b. If f and H are in the same units, the ratio $\frac{f}{H}$ is called a 1-1 scale factor. Determine the 1-1 scale factor for a photograph taken at a height of 30 km with a camera having a 150 mm focal length.

Solution: $\frac{f}{H} = \frac{150 \times 10^{-3}}{30 \times 10^{3}} = 5 \times 10^{-6}$

c. If the photograph of part (a) shows an image of a straight road that measures 1.25 mm on the film, how long is the actual road?

Solution: Let L be the actual length of the road in meters.

$$\frac{\text{image length}}{\text{actual length}} = \frac{1.25 \times 10^{-3}}{L} = 5 \times 10^{-6}$$

$$L = \frac{1.25 \times 10^{-3}}{5 \times 10^{-6}} \text{ m} = 0.25 \times 10^{3} \text{ m} = 250 \text{ m}$$

d. With current technology, it is possible to make measurements on photographs to the nearest micron (10^{-6} m). What is the smallest actual length whose image can be measured on the photograph of part (b)? (This is called the resolution of the photograph.)

Solution: Let S be the smallest actual length in meters. Then

$$\frac{10^{-6}}{S} = 5 \times 10^{-6},$$

so

$$5 \times 10^{-6}\, S = 10^{-6}$$

and

$$S = \frac{1}{5}\,\text{m} = 0.2\,\text{m}.$$

The curvature of Earth can introduce distortion in a photograph. We need some new terminology to discuss the correction for such distortion. The point on Earth vertically below the camera is called the *nadir* and its image on the film is called the *photograph nadir point.* Because the photograph is flat, the image of any point except the nadir will be closer to the photograph nadir point than it would be if Earth were also flat. In the next problem, we develop a formula to correct for this.

PROBLEM 8. The geometry of the photographic correction for Earth's curvature is shown in Fig. 4.9. The image of the point P is a distance r from the photograph nadir point Q, f is the focal length of the camera, and H is its height when the picture was taken. In order to get a "corrected" picture, we need to place the image at P′, in the plane of the tangent to the nadir, N. This means we need to compute Δr so that the corrected image is a distance $r + \Delta r$ from Q.

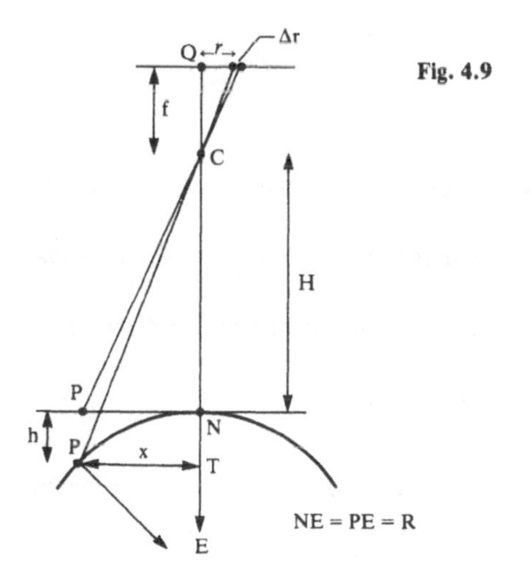

Fig. 4.9

a. Show that $\Delta r \doteq \dfrac{Hr^3}{2Rf^2}$, where R is the radius of Earth.

Solution: Let h be the vertical displacement of P′ with respect to P and let x be the horizontal displacement of P′ (also of P) with respect to N. We see from the diagram in Fig. 4.9 that x and h are related. If T is the foot of the perpendicular from P to NE, where E is the center of Earth, then ΔPTE is a right triangle with PE = R, PT = x, and TE = $R - h$. By the Pythagorean theorem, $R^2 = x^2 + (R - h)^2$, so $R^2 = x^2 + R^2 - 2Rh + h^2$, giving $x^2 = 2Rh - h^2$. Since h is very small compared to R, we shall use the approximation $x^2 \doteq 2Rh$.

There are two pairs of similar triangles in the diagram.

For the inner pair,

$$\frac{f}{r} = \frac{H + h}{x},$$

or

$$fx = (H + h)r,$$

or

$$fx = Hr + hr.$$

For the outer pair,

$$\frac{f}{r + \Delta r} = \frac{H}{x},$$

or

$$fx = H(r + \Delta r),$$

or

$$fx = Hr + H\Delta r.$$

Comparing the two expressions for fx, we see that $hr = H\Delta r$. Then $\Delta r = \frac{r}{H}h \doteq \frac{r}{H}\frac{x^2}{2R}$. Now we need another approximation in order to eliminate x. In the relation $fx = Hr + hr$, since h is small compared to H, we have $fx \doteq Hr$, so $x^2 = \frac{H^2r^2}{f^2}$. Making this last substitution, we have $\Delta r = \frac{r}{2HR} \cdot \frac{H^2r^2}{f^2} = \frac{Hr^3}{2Rf^2}$.

b. Find the correction Δr and the resulting $r + \Delta r$ for a photographic image taken at a height of 92 km with a camera having a focal length of 132 mm if r measures 65.24 mm. Recall that $R = 6400$ km.

Solution: Since H and R are in km and r and f are in mm, if we do no unit conversions, we shall be computing Δr in mm.

$$\Delta r = \frac{(92)(65.24)^2}{(2)(6400)(132)^2} = 0.11 \text{ mm}$$

$$r + \Delta r = 65.24 + 0.11 = 65.35 \text{ mm}$$

The final problems in this chapter deal with some aspects of planning for the future construction of such large commercial space structures as the antenna system in the illustration on page 47. It is planned that the materials for the antenna system will be carried up and the actual construction done in orbit. This frees the construction of two considerations: (1) the rigidity that would be required for such a structure to break away from Earth's gravity and (2) the strength needed to survive transportation into orbit intact. It is, of course, desirable to keep to a minimum the number of trips needed to transport the components, and considerable effort has gone into the development of materials that are strong and lightweight and that maintain their properties over a wide range of temperatures. Let us see how successful the effort to minimize the number of trips has been.

PROBLEM 9. The Space Shuttle can carry 29 500 kg of payload into orbit in a cargo bay that is basically a cylinder having a length of 18.3 m and a diameter of 4.6 m. The structure in the illustration has 91 antennas, each a paraboloidal cap 20 m in diameter and 2 m deep. The material for the antennas is a knitted metallic mesh weighing 60 g/m^2.

The plan for the truss assembly that holds the antennas is shown in Fig. 4.10. A promising material for the columns is graphite-epoxy, which combines excellent strength and stiffness with light weight, having a density of 1522 kg/m^3. The truss assembly shown has 252 copies of the basic repeating element, with each repeating element consisting of a tetrahedron having nine complete struts as shown in Fig. 4.10(c). The struts themselves are hollow columns 10.4 m long with radius 3.8 cm and thickness 0.57 mm as shown in Fig. 11.

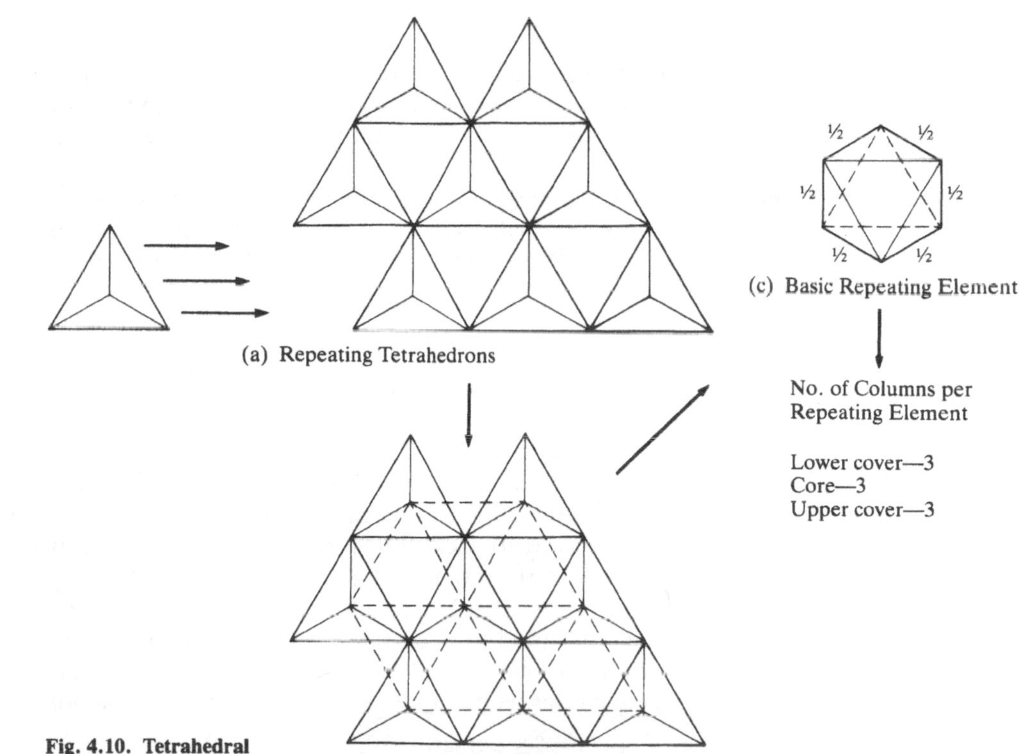

Fig. 4.10. Tetrahedral truss construction.

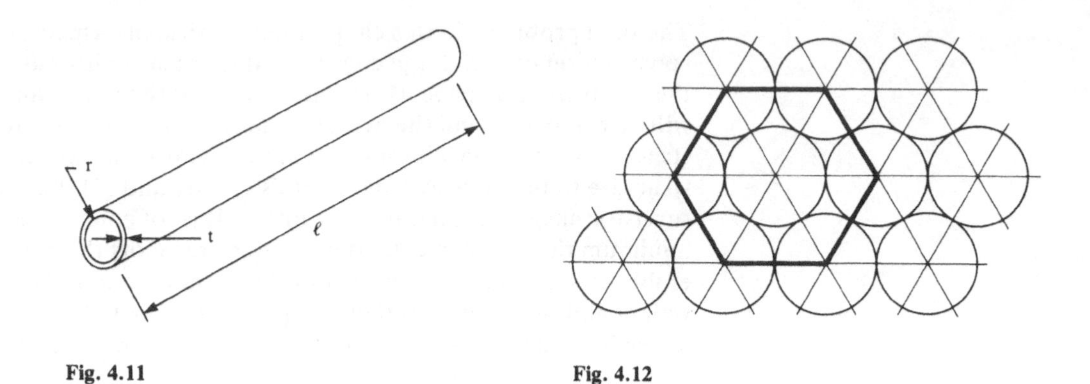

Fig. 4.11

Fig. 4.12

a. How many Shuttle trips would be necessary to get the weight of these elements (the metallic mesh for the antennas and the columns for the struts) into orbit?

Solution: We shall get an approximation for the weight of metallic mesh by treating the antennas as though they were circles of radius 10 m. (In Chapter 10, "Calculus," we shall get a more accurate result.) If we have 91 circles of radius 10 m, the total area will be $(91)(\pi)(10)^2$ m^2, and the total weight of metallic mesh will be $(91)(\pi)(10)^2(60)$ g $= 1.7 \times 10^3$ kg. The volume of material in each column can be approximated by $(2\pi r)(t)(\ell) = 2\pi(3.8 \times 10^{-2})(0.57 \times 10^{-3})(10.4)$ m^3, so the total weight of the columns is $(252)(9)(6.28)(3.8)(0.57)(10.4)(10^{-5})(1522)$ kg $=$ 4.9×10^3 kg. The total weight of these materials is $(1.7 + 4.9) \times 10^3$ kg $= 6.6 \times 10^3$ kg. $\frac{6.6 \times 10^3 \text{ kg}}{29.5 \times 10^3 \text{ kg}} = 0.224$, or about 22 percent of the Shuttle's weight capacity.

b. We see that the Shuttle can easily carry this weight on a single trip. Now we must consider volume: will the materials fit in the available space? Assume that the metallic mesh is 7.5 mm thick and sufficiently flexible to pack into any shape.

Solution: The cargo bay's cylindrical volume is given by $\pi R^2 L = \pi(2.3)^2(18.3)$ m$^3 = 3.0 \times 10^2$ m^3. We have already found that the total area of metallic mesh is $(91)(100\pi)$ m^2 $= 2.9 \times 10^4$ m^2, so the total volume of mesh is $(2.9 \times 10^4)(7.5 \times 10^{-3})$ m^3 $= 2.2 \times 10^2$ m^3.

This leaves $(3.0 - 2.2) \times 10^2$ m$^3 = 80$ m^3 for the columns (and all the remaining hardware needed for assembly, which we are ignoring here).

Since the columns are 10.4 m long, they cannot be placed end to end in the 18.3 m long cargo bay. We must consider how to stack them most efficiently. If we consider the cross section of the stack, we see that we need to find the most efficient way to pack circles in a plane. It is intuitively reasonable (although the proof is far from simple) that the maximum efficiency is achieved when each circle is tangent to six others, as illustrated in Fig. 4.12. From the diagram we see that each hexagon has sides of length $2r$, where r is the radius of the circle, and therefore has area $6(1/2)(2r)(\sqrt{3}r) = 6\sqrt{3}r^2$. Also, each hexagon contains three complete circles whose total area is $3(\pi r^2)$. So the fraction of area occupied by circles is $\frac{3\pi r^2}{6\sqrt{3}r^2} = \frac{\pi}{2\sqrt{3}} = 0.907$.

Since the fraction of space occupied at the boundary will be smaller than this, let us assume that the columns pack into the cargo bay so that 12 percent of the space is empty. This means that the columns will pack into a cross-sectional area equal to 1/0.88 of their total cross-sectional area. We recall that there were $252 \times 9 = 2268$ columns, each having radius 3.8 cm and therefore a cross-sectional area of $\pi(0.038)^2 \text{ m}^2 = 4.5 \times 10^{-3} \text{ m}^2$. The total cross-sectional area needed, then, is $\frac{2268 \times 4.5 \times 10^{-3}}{0.88} \text{ m}^2 = 11.7 \text{ m}^2$. Since the columns are 10.4 m long, they will occupy $(11.7)(10.4) \text{ m}^3 = 1.2 \times 10^2 \text{ m}^3$ of space. However, there was only 80 m^3 of space left after calculating the volume of the metallic mesh, so it will take more than one trip to handle the volume, even though the weight is not a problem. Our success in reducing the weight now places the focus of our attention on volume.

PROBLEM 10. In order to fit more columns into a smaller space, the designers realized that they should investigate the possibility of tapering the columns and then "nesting" them for transportation, like a stack of paper cups. Fig. 4.13 illustrates the idea. Under this scheme, each column would be made of two tapered half-columns, with their wider openings joined; half-columns could then be nested for stowage in the cargo bay. Tapered columns have been developed and tested for strength. If r_1 is the radius of the smaller end and r_2 the radius of the larger end, tests showed that an optimum taper ratio is $\frac{r_1}{r_2} = 0.41$ and that such a tapered column is actually stronger; it can carry about 30 percent more load before buckling than an untapered column of the same weight.

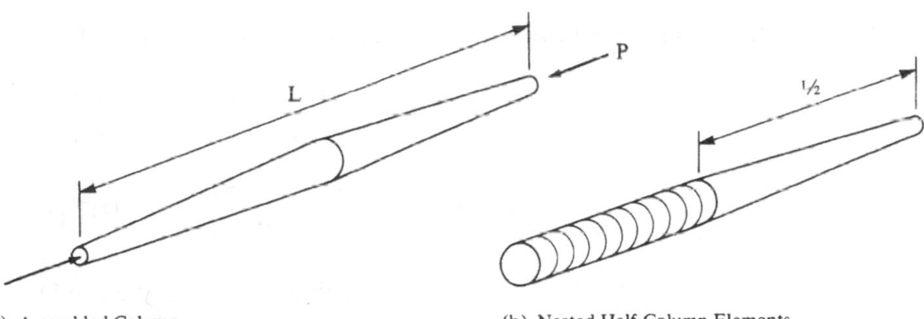

(a) Assembled Column (b) Nested Half-Column Elements

Fig. 4.13. Tapered Column Concept

a. If the mean radius is to be 3.8 cm as before, and $\frac{r_1}{r_2} = 0.4$, find the values of r_1 and r_2.

Solution: We have $\frac{r_1 + r_2}{2} = 3.8$ and $r_1 = 0.4\, r_2$. Clearing the fraction and substituting, we get

$$0.4\, r_2 + r_2 = 7.6$$

$$1.4\, r_2 = 7.6$$

$$r_2 = \frac{7.6}{1.4} = 5.4 \text{ cm}$$

$$r_1 = (0.4)(5.4) = 2.2 \text{ cm}.$$

b. Fig. 4.14(a) and (b) display the geometry of the tube nesting, where $d_1 = 2 \cdot r_1$, $d_2 = 2 \cdot r_2$, ℓ is the length of a half-column, and Δ is the tube-nesting separation. Show that $\Delta \doteq \frac{t\ell}{r_2 - r_1}$ and find an expression in terms of ℓ and Δ for the number of half-columns that will fit into one stack the length of the Space Shuttle cargo bay.

Solution: In Fig. 4.14(b), if we insert the horizontal line shown and letter some key points as indicated (Fig. 4.15), $\Delta ABE'$ and ΔBCD are similar, so $AB/BC = BE'/CD$. We have $AB = \Delta$, $BC = \ell$, $CD = r_2 - r_1$, and we shall approximate BE' by $BE = t$. Then, with this approximation and the proportion above, $\Delta \doteq \frac{t\ell}{r_2 - r_1}$.

From Fig. 4.14(a), we have one half-column of length ℓ on the left end, in which we nest n additional half-columns, where $n = \text{INT}\left[\frac{18.3 - \ell}{\Delta}\right]$. (INT signifies the greater integer which is less than or equal to the number in the square bracket.) Now the number of half-columns that will fit into one stack is $N = 1 + n = 1 + \text{INT}\left[\frac{18.3 - \ell}{\Delta}\right]$.

c. For the truss assembly of Problem 9, determine the volume occupied by the strut columns if they are made of half-columns as described here and nested for stowing in the cargo bay.

Solution: We have ℓ = half-column length $= (1/2)(10.4) = 5.2$ m

$$\Delta \doteq \frac{t\ell}{r_2 - r_1} = \frac{(0.57 \times 10^{-3})(5.2)}{(5.4 - 2.2) \times 10^{-2}} \text{ m} = 9 \times 10^{-2} \text{ m}$$

$$N = 1 + \text{INT}\left[\frac{18.3 - 5.2}{9 \times 10^{-2}}\right] = 1 + \text{INT}[145.6] = 1 + 145 = 146.$$

We had a total of 2268 columns, or 4536 half-columns, so this means there will be $\text{INT}\left[\frac{4536}{146}\right] = 31$ stacks, and one additional shorter stack.

Each stack is 18.3 m long (although one stack will be shorter) and has a radius of 5.4 cm, so its volume is $\pi(0.054)^2(18.3) = 0.17 \text{ m}^3$. The total volume of the 32 stacks is a little less than $32 \times 0.17 = 5.4 \text{ m}^3$.

By the analysis in the last part of Problem 9, these stacks will take up $\frac{5.4}{0.88} = 6.2 \text{ m}^3$ of space in the cargo bay, and now the materials for the truss assembly and the antenna "dishes" can all be transported in a single Shuttle trip.

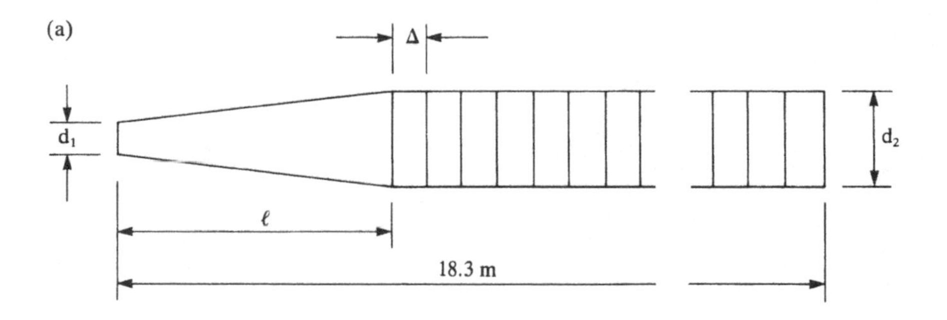

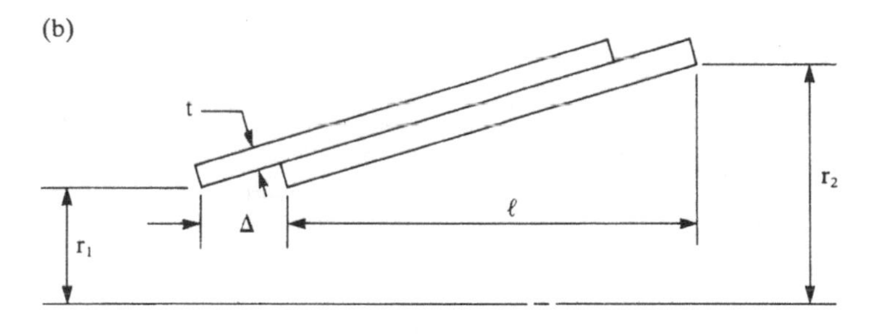

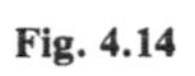

Fig. 4.14

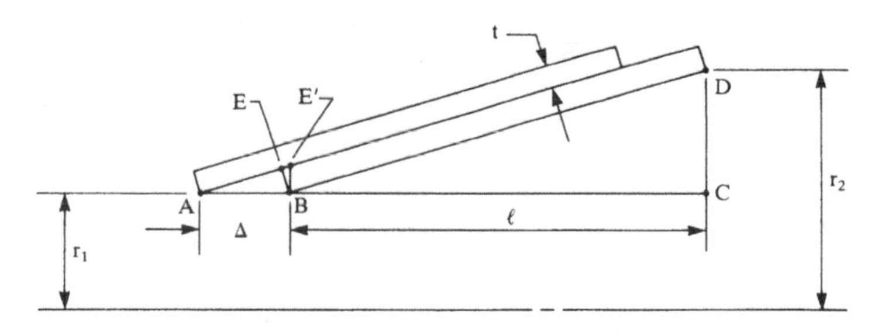

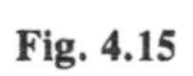

Fig. 4.15

CHAPTER FIVE

PROBABILITY AND STATISTICS

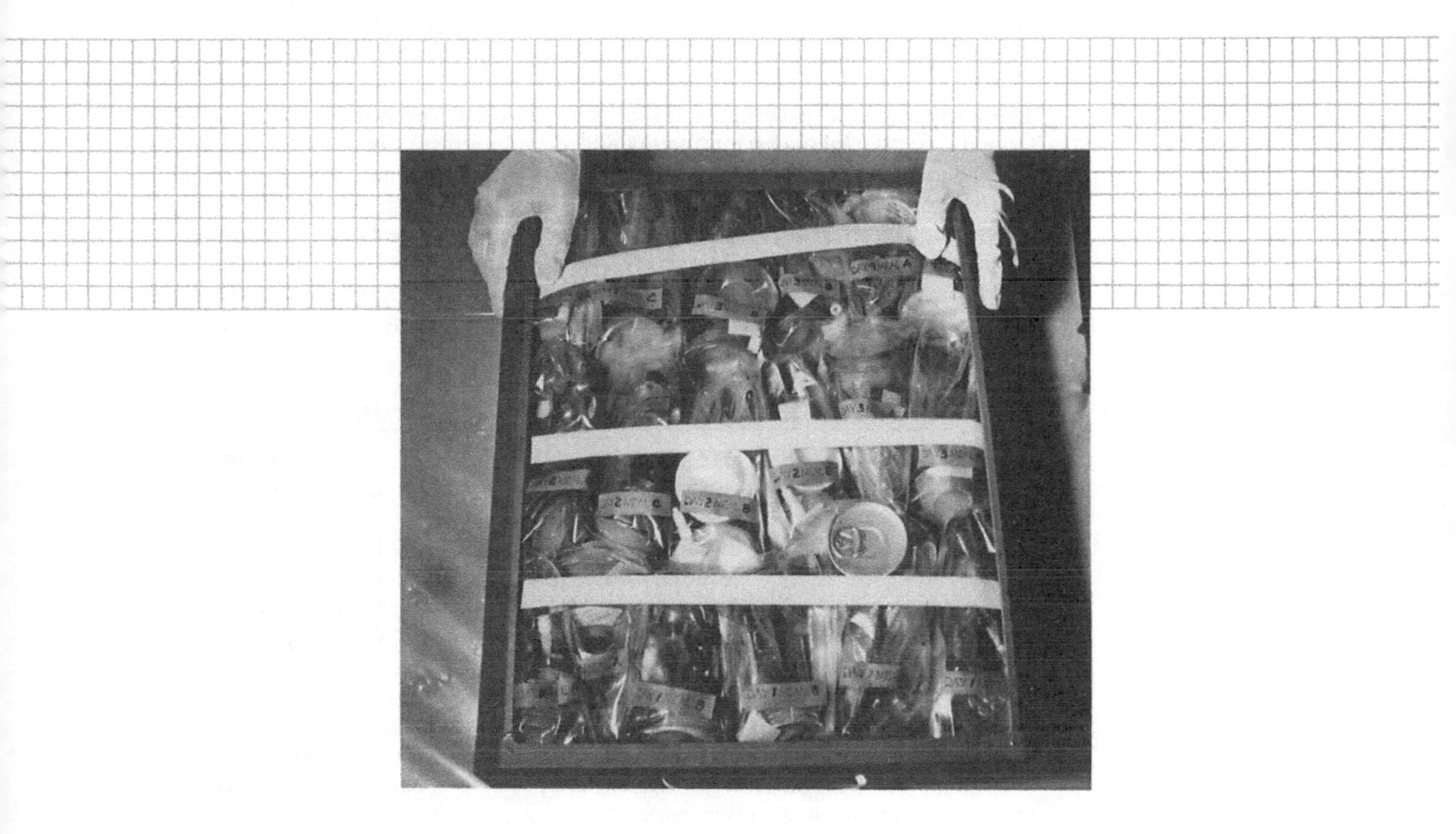

Space Shuttle locker tray containing 18 standard orbital flight test menu meals for two crew members.

Probability theory and statistical techniques make important contributions to the space program. In this chapter we examine the role of probability in menu planning, in some aspects of the transmission and coding of spacecraft observations, and in the control of equipment reliability. Some elementary examples of the use of statistics are illustrated in the final two problems.

PROBLEM 1. The early manned spaceflights revealed much about the body's response to prolonged weightlessness. An interesting and varied food supply was thus needed to guard against a loss of appetite in the face of what was learned. The food supply for the crew of the Space Shuttle is carefully planned to compensate for the high energy requirements (averaging 3000 calories per person per day) of working in a frictionless environment and the body's tendency to lose essential minerals (such as potassium, calcium, and nitrogen) in microgravity. The Space Shuttle food and beverage list contains more than a hundred individual items. A typical day's menu might be the following:

Meal I	*Meal II*	*Meal III*
Peaches	Frankfurters	Shrimp cocktail
Beef patty	Turkey tetrazzini	Beef steak
Scrambled eggs	Bread (2)	Rice pilaf
Bran flakes	Bananas	Broccoli au gratin
Cocoa	Almond crunch bar	Fruit cocktail
Orange drink	Apple drink (2)	Butterscotch pudding
		Grape drink

In general, each meal III contains a main dish, a vegetable, and two desserts, with an appetizer included every other day. The food list contains 10 items classified as main dishes (M), 8 vegetable dishes (V), 13 desserts (D), and 3 appetizers (A). How many different menu combinations are possible in each of the first six days of flight, assuming no dish is repeated?

Solution: The number of choices is tabulated below:

Day	*A*	*M*	*V*	*D1*	*D2*	*Number of combinations*
1	3	10	8	13	12	37 440
2	—	9	7	11	10	6 930
3	2	8	6	9	8	6 912
4	—	7	5	7	6	1 470
5	1	6	4	5	4	480
6	—	5	3	3	2	90

PROBLEM 2. The electronic telemetry system aboard a spacecraft transmits the data of spacecraft motion in the x, y, and z directions. The system consists of three motion sensors, a signal conditioner, and a transmitter. The probability of failure for each motion sensor and for the signal conditioner is 0.0001. The probability of failure for the transmitter is 0.001. Assuming that component failures are independent events and that the failure of any component will render the telemetry system inoperative, compute the probability of a spacecraft telemetry success.

Solution: The probability of success is equal to 1 minus the probability of failure. Therefore, the probability of success for each sensor and the signal conditioner is

$$P = 1 - 0.0001 = 0.9999.$$

Similarly, the probability of success for the transmitter is

$$P = 1 - 0.001 = 0.999.$$

The probability of success for the telemetry system is the product of probabilities of success for each component; that is,

$$P = (0.9999)^4 (0.999) = 0.9986.$$

The signals transmitted by a spacecraft telemetry system are in the form of pulses imposed on a radio beam, which can be interpreted as binary digits. For example, the signal fragment . . . ⎍⎍ . . . will be read as . . . 010110 . . . , since the presence of a pulse is read as 1 and the absence of a pulse as 0. Each possible representation of a 0 or a 1 is called a "bit."

For a variety of reasons, equipment errors can cause a 0 to be transmitted instead of a 1, or vice versa. As a result, error-detecting codes have been developed to improve data reliability. All such codes are based on transmitting extra bits that can be used to determine whether errors are present and even, for the more sophisticated codes, where the errors are. Transmitting these extra bits, however, means that fewer message-carrying bits can be sent in a given unit of time, and so transmission reliability must be traded against transmission efficiency. Probability theory plays an important part in weighing the trade-offs.

PROBLEM 3. a) The telemetry system of a certain spacecraft has a probability of 1 percent of transmitting an erroneous bit. One way to increase data reliability would be to repeat each message bit three times. For example, . . . 010110 . . . would become . . . 000111000111111000 . . . , if no errors occur. If it is decided to interpret any of the triplets 000, 001, 010, or 100 as 0 and any of the triplets 011, 101, 110, or 111 as 1, find the probability of error in the interpretation of a message bit, assuming the transmission of each bit is independent.

Solution: A message bit will be interpreted erroneously if two or three errors have occurred in the triplet.

$$P(2 \text{ errors}) = \binom{3}{2} (0.01)^2 (0.99) = 0.000297$$

$$P(3 \text{ errors}) = (0.01)^3 = 0.000001$$

$$P(2 \text{ or } 3 \text{ errors}) = 0.000298 \doteq 0.0003$$

We see that we can reduce the probability of a transmission error in a single bit from 1 percent to 0.03 percent, but at a cost of sending three times as many bits as are actually needed for the message. To put it a different way, the desired message would be sent one-third as quickly.

b. More efficient error detection can be done with *parity coding*. In this method, a "parity bit" is inserted after each string of message bits of a predetermined length k so that the sum of the $(k + 1)$ bits is either always even (even parity) or always odd (odd parity). For example, if $k = 4$ and even parity is used, the message 110100101001 . . . will become 110110010110010 . . . On receiving the transmission signals, an error is detected if the sum of the appropriate five contiguous digits is odd. If the probability of error in a single bit is 1 percent, find (i) the probability of at least one error in the transmission of four sequential bits, and (ii) the probability of an undetected error after using even-parity coding.

Solution: (i) The probability of an error occurring in at least one of the four bits is

$$1 - P(\text{no errors in the 4 bits}) = 1 - (0.99)^4$$

$$= 1 - 0.9606 \doteq 0.04, \text{ or 4 percent.}$$

(ii) In each set of five bits under parity coding, if 1, 3, or 5 errors occur, the sum of the binary digits will be odd and the error will be detected. If 2 or 4 digits are in error, this will go undetected.

$$P(2 \text{ errors}) = \binom{5}{2}(0.01)^2(0.99)^3 \doteq 0.00097$$

$$P(4 \text{ errors}) = \binom{5}{4}(0.01)^4(0.99) \doteq 5 \times 10^{-8}$$

so

$$P(\text{undetected error}) = P(2 \text{ or } 4 \text{ errors}) = 0.1 \text{ percent}$$

By inserting a parity bit after each four message bits, we have reduced the transmission efficiency to 80 percent of its possible maximum but have reduced the probability of an undetected transmission error in each four-bit "word" from 4 percent to 0.1 percent. However, when we do detect an error, parity coding does not tell us which of the bits is erroneous. In Chapter 8, "Matrix Algebra," we shall examine the Hamming Code, which not only detects a transmission error but also tells which bit is wrong.

PROBLEM 4. An aerospace consulting company is working on the design of a spacecraft system composed of three main subsystems, A, B, and C. The reliability, or probability of success, of each subsystem after three periods of operation is displayed in the following table:

	1 day	*3.3 months*	*8.5 months*
A	0.9997	0.8985	0.6910
B	1.0000	0.9386	0.7265
C	0.9961	0.9960	0.9959

These reliabilities have been rounded to four significant digits. The 1.0000 in the first column means that the likelihood of the failure of subsystem B during the first day of operation is so remote that more than four significant digits are needed to indicate it.

a. Consider the case of the series system shown in Fig. 5.1. If any one (or more) of the subsystems A, B, or C fails, the entire system will fail. If P_s is the total probability of success of the system, find P_s for each of the three time periods.

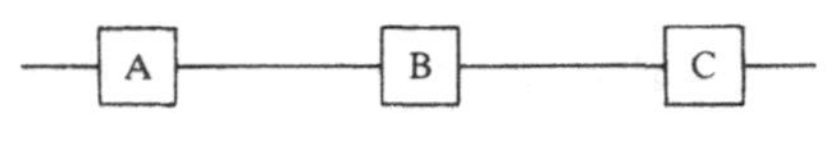

Fig. 5.1

Solution: For the first 24 hours,

$$P_s = P_A P_B P_C$$
$$= (0.9997)(1.0000)(0.9961)$$
$$= 0.9958.$$

For a period of 3.3 months,

$$P_s = P_A P_B P_C$$
$$= (0.8985)(0.9386)(0.9960)$$
$$= 0.8400.$$

For a period of 8.5 months,

$$P_s = P_A P_B P_C$$
$$= (0.6910)(0.7265)(0.9959)$$
$$= 0.5000.$$

b. The system shown in Fig. 5.2 will succeed if B succeeds and at least one of A and C succeeds. Find the probability of success for this system for the 3.3-month time period.

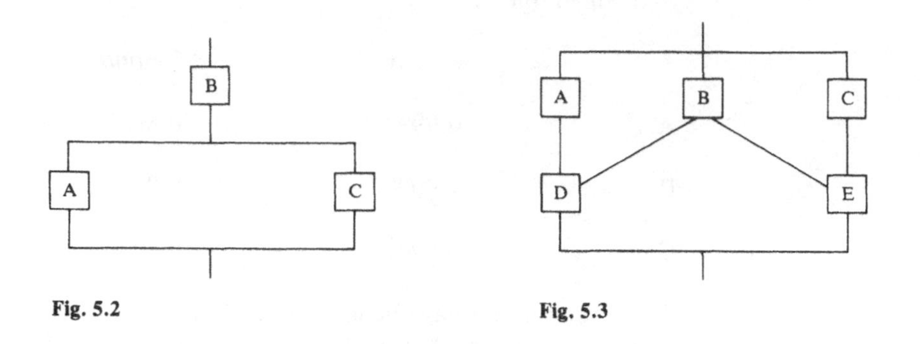

Fig. 5.2

Fig. 5.3

Solution: The probability of success for the portion of the system containing A and C is

$$P_{A,C} = 1 - P(\text{both A and C fail})$$
$$= 1 - (0.1015)(0.0040) = 0.9996.$$

Then

$$P_s = P_B P_{A,C} = (0.9386)(0.9996) = 0.9382.$$

c. For more complicated systems, the use of conditional probability is helpful. If an event A can be divided into n mutually exclusive subevents $B_1, B_2, \ldots B_n$ (n finite), then $P(A) = P(A|B_1)P(B_1) + P(A|B_2)P(B_2) + \ldots + P(A|B_n)P(B_n)$, where the notation $P(X|Y)$ designates the conditional probability of X given that Y has occurred.

Consider the system in Fig. 5.3, where the 3.3-month reliabilities of the subsystems A, B, C are the same as before and the 3.3-month reliabilities of D and E are 0.9216 and 0.9542, respectively. Use conditional probability to find the reliability (i .e., P_s) of this system for the 3.3-month period.

Solution: This system will succeed if any one of the paths (A,D), (B,D), (B,E), or (C,E) succeeds. We can choose B as our focus and assert that $P_s = P(\text{system succeeds}|\text{B succeeds})\, P_B + P(\text{system succeeds}|\text{B fails})\,(1 - P_B)$. Now we evaluate $P(\text{system succeeds}|\text{B succeeds})$. If the system succeeds given that B succeeds, this means that at least one of D and E would have succeeded, so

$$P(\text{system succeeds}|\text{B succeeds}) = 1 - (1 - P_D)(1 - P_E)$$
$$= 1 - (1 - P_D - P_E + P_D P_E)$$
$$= P_D + P_E - P_D P_E$$
$$= (0.9216) + (0.9542) - (0.9216)(0.9542)$$
$$= 0.9964.$$

Next we evaluate P(system succeeds|B fails). For the system to succeed in view of the failure of B means that at least one of the paths (A,D) or (C,E) must have succeeded, so

$$P(\text{system succeeds}|\text{B fails}) = 1 - (1 - P_A P_D)(1 - P_C P_E)$$

$$= P_A P_D + P_C P_E - P_A P_D P_C P_E$$

$$= (0.8985)(0.9216) + (0.9960)(0.9542)$$

$$- (0.8985)(0.9216)(0.9960)(0.9542)$$

$$= 0.9915$$

$$P_s = (0.9964)P_B + (0.9915)(1 - P_B)$$

$$= (0.9964)(0.9386) + (0.9915)(0.0614)$$

$$= 0.9961.$$

PROBLEM 5. In Problem 4a we saw that the total reliability of the system deteriorates rather rapidly in its present stage of design, with less than a 50-percent chance that it will operate after 8.5 months. The reliability of subsystem C remains nearly constant, whereas the greatest decline in reliability takes place in subsystem A, which contains a particular part that is expected to wear out rapidly. The consulting firm is asked to determine if enough improvement could be made in subsystem A to provide a reliability after 8.5 months of 0.7500. Compute the improvement needed in subsystem A.

Solution: Let x be the factor by which the reliability of subsystem A must be multiplied. Then, as before,

$$P_s = P_A P_B P_C$$

$$0.7500 = (0.6910x)(0.7265)(0.9959) = 0.5000x$$

$$x = \frac{0.7500}{0.5000} = 1.500.$$

The reliability of subsystem A must be $1.500 \times 0.6910 = 1.037$. The increase in reliability cannot be obtained by improving subsystem A alone, since the reliability cannot be greater than 1.

The next problem demonstrates the combined use of probability and computer simulation to determine the volume of an irregular solid.

PROBLEM 6. An internal fuel tank on a space vehicle has the shape of an ellipsoid truncated by three planes, as shown in Fig. 5.4. Our problem is to determine the volume of this fuel tank. Let us use, for an example, the ellipsoid whose equation is $\frac{x^2}{8^2}+\frac{y^2}{3^2}+\frac{z^2}{2^2}=1$ and with the planes being $x = \pm 7$, and $z = -1.5$, where the units are meters.

Fig. 5.4

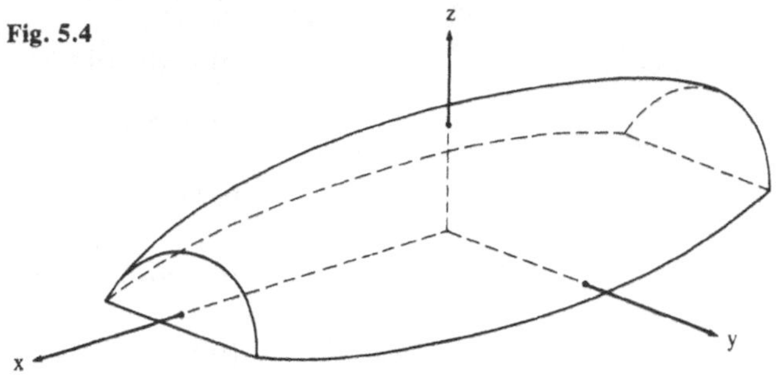

a. If the tank is surrounded as tightly as possible by a rectangular prism with faces parallel to the planes formed by the coordinate axes, what inequalities must the coordinates of the ponts inside the prism satisfy? What is the volume of this prism?

Solution: If (x, y, z) is inside the prism, x must satisfy $-7 < x < 7$ because of the truncating planes $x = -7$ and $x = 7$; y must satisfy $-3 < y < 3$ because $y = -3$ and $y = 3$ are the planes tangent to the ellipsoid and parallel to the x-z plane; z must satisfy $-1.5 < z < 2$, since z is bounded below by the truncating plane $z = -1.5$ and above by the plane $z = 2$ tangent to the ellipsoid and parallel to the x-y plane. This rectangular prism has dimension 14 m × 6 m × 3.5 m, and the resulting volume is 294 m^3.

b. Let V_p be the volume of the prism and let V_t be the volume of the tank, which we are seeking. If a point is randomly chosen inside the prism, express the probability that it is also inside the tank, in terms of V_p and V_t.

Solution: This probability is equal to V_t/V_p, the ratio of the volume of the tank to that of the surrounding prism.

c. If N points are chosen at random inside the prism and I of these points are also inside the tank, express V_t in terms of N, I, and V_p.

Solution: The probability that I points are in the tank out of the N points chosen randomly inside the prism is approximated by I/N. So we get $I/N = V_t/V_p$, giving $V_t = V_p(I/N)$.

d. Write a computer program to perform this simulation, using a random number generator to get coordinates of points within the prism.

Solution:

```
10  REM  VOLUME SIMULATION USING
      PROBABILITY
12  DIM K(25): DIM VOL(25): DIM P
     CT(25)
15 M = 1
20  REM  INTERNAL FUEL TANK, TRUN
     CATED   ELLIPSOID
30  REM  BOUNDARIES X^2/8^2 + Y^2
      /3^2 +  Z^2/2^2 = 1,X= -7,X=7
     ,7=1,5
32  PRINT "THIS PROGRAM COMPUTES
      THE VOLUME OF "
34  PRINT "A SPACECRAFT"S INTERNA
     L FUEL TANK "
36  PRINT "WHOSE SHAPE IS A TRUNC
     ATED ELLIPSOID"
37  PRINT "WITH BOUNDARIES Z= -1.5
     ,X= -7,X=7,  AND"
38  PRINT "X^2/8^2 PLUS Y^2/3^2 + Z^
     2/2^ =1"
39  PRINT "THE USER WILL CHOOSE T
      HE NUMBER OF": PRINT "POINTS
       IN THE SIMULATION": PRINT
40  PRINT "HOW MANY POINTS DO YOU
     CHOOSE?"
42  INPUT K(M)
50  PRINT "COMPUTING..."
80 VP=14  * 6 * 3.5
90 I = 0
100  FOR N = 1 TO K(M)
110 X =   - 7 + 14 *  RND (1)
120 Y =   - 3 + 6 *  RND  (1)
130 Z =   - 1.5 + 3.5 *  RND (1)
140  GOSUB 1000
150  NEXT N
160 VOL(M) =  VP * I / K(M)
161 X% = VOL(M):VOL(M) = X%
165 PCT(M) =  100 * I / K(M)
166 P% =  PCT(M):PCT(M) P%
170  PRINT I;" OF THE ";K(M);" PO
     INTS"
180  PRINT "WERE IN THE TANK,FOR
     A VOLUME ESTIMATE"
185  PRINT VOL(M);" CUBIC UNITS."

190  PRINT " THE PERCENTAGE OF PO
     INTS IN THE TANK"
200  PRINT "WAS ";PCT(M)
202  GOSUB 1500
205  PRINT : PRINT "* * * * * AND
     THER TRIAL? * * * * *"
210  PRINT : PRINT : PRINT "IF YO
     U WANT TO REDO THE SIMULATIO
     N"
220  PRINT "WITH THE SAME NUMBER
     OF POINTS,TYPE S"
230  PRINT : PRINT "WITH A DIFFER
     ENT NUMBER,TYPE D"
240  PRINT : PRINT : PRINT "TYPE
     ANY OTHER LETTER IF FINISHED
250  INPUT A$:M=M + 1
260  IF A$ =  "S" THEN K(M) =  K(M -
     1): GOTO 50
270  IF A$ =  "D" THEN 40
280  END
1000  REM  TEST POINT
1010 P =  X * X / 64 + Y * Y / 9 +
     Z * Z / 4
1020  IF P > = 1 THEN 1040
1030 I = I + 1
1040  RETURN
1500  REM SUMMARY OF TRIALS
1510  PRINT : PRINT " * * * * * S
     UMMARY? * * * * *"
1520  PRINT : PRINT "DO YOU WANT
     TO SEE A SUMMARY OF THE"
1525  PRINT "SIMULATIONS SO FAR?"
     : INPUT B$
1530  IF B$ <  > "Y" THEN 1600
1540  PRINT : PRINT : PRINT "
     N            PCT            VOL"
1545  PRINT
1550  FOR J =  1 TO M
1560  HTAB (5): PRINT K(J);
1570  HTAB (16): PRINT PCT(J);
1580  HTAB (25): PRINT VOL(J)
1590  NEXT J
1600  RETURN


]RUN
THIS PROGRAM COMPUTES THE VOLUME OF
A SPACECRAFT'S INTERNAL FUEL TANK
WHPSE SHAPE IS A TRUNCATED ELLIPSOID
WITH BOUNDARIES Z= -1.5,X= -7,X=7,
AND X^2/8^2 + Y^2/3^2 + Z^2/2^2 =1
THE USER WILL CHOOSE THE NUMBER OF
POINTS IN THE SIMULATION


 * * * * * SUMMARY? * * * * *

DO YOU WANT TO SEE A SUMMARY OF THE
SIMULATIONS SO FAR?
?Y

     N            PCT            VOL

   100            71            208
   100            61            179
   200            62            183
   200            64            189
   400            62            184
```

PROBLEM 7. Sunspots were observed and recorded as long as two thousand years ago. The invention of the telescope around 1610 permitted the systematic observation of these solar features, their motion, and their frequency of occurrence. (Problem 6 of Chapter 7 illustrates the use of trigonometry in analyzing sunspot motion.) It is relatively easy to observe sunspots by using a long-focus telescope to project an image of the Sun on a piece of white cardboard.

Fig. 5.5 shows the record from 1610 to 1975 of what is now commonly referred to as the "sunspot cycle." The vertical scale represents the number of sunspots observed. The data since 1740 are considered reliable.

Although sunspots are still not well understood, it has been established that they are regions in the solar atmosphere that contain enormous magnetic fields relative to their surroundings, along with cooler temperatures. Moreover, there appear to be connections between the level of sunspot activity and the occurrence of "magnetic storms" in Earth's ionosphere, the density of Earth's upper atmosphere, and changes in Earth's weather and climate.

Since variations in upper atmosphere density can affect the orbital lifetimes of satellites, the prediction of sunspot activity is an important aspect of the planning of some space missions. The mean cycle length as well as its variability must be taken into account, making statistical analysis vital to such predictions.

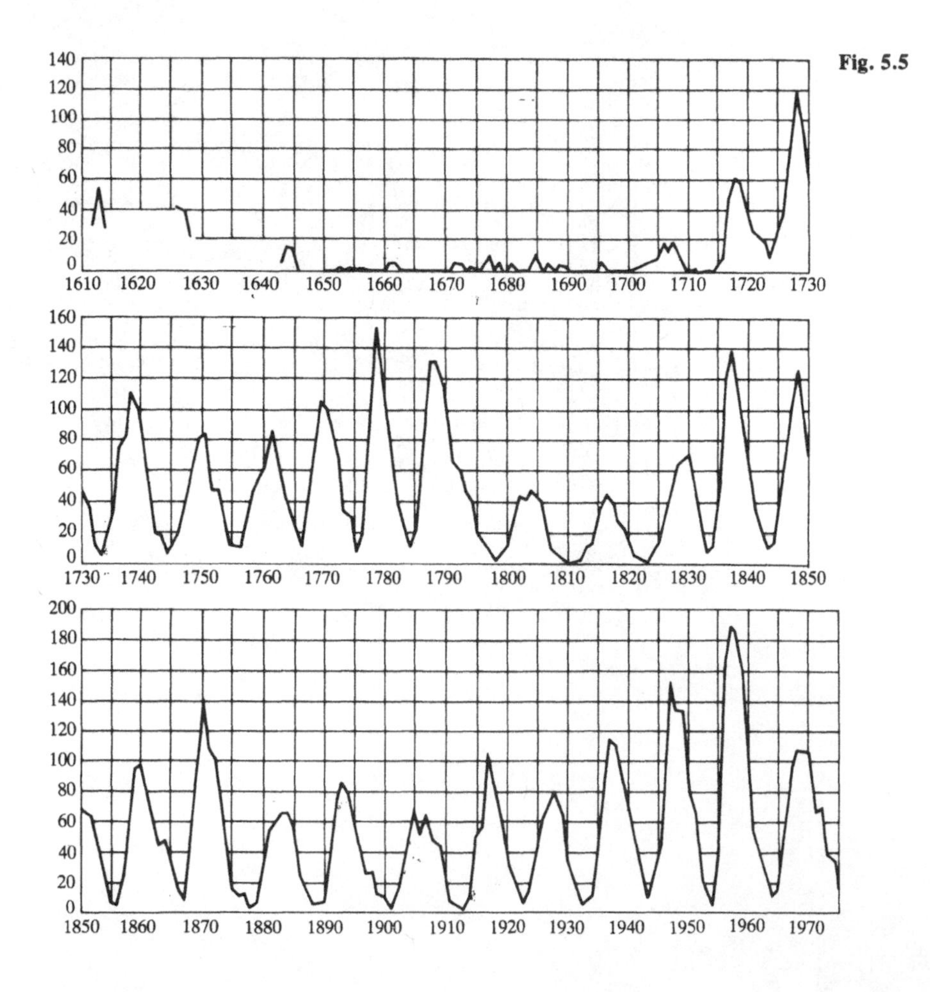

Fig. 5.5

The following table summarizes some of the data of Fig. 5.5. The first step in the statistical analysis for the prediction of sunspot activity is to determine the mean and the standard deviation for each of the following measures: the rise time; the fall time; the period from minimum to minimum; the period from maximum to maximum. Compute these means and standard deviations.

Table 5.1

Cycle	Year of Minimum	Year of Maximum	Cycle	Year of Minimum	Year of Maximum
1	1745	1750.3	12	1867.2	1870.6
2	1755.2	1761.5	13	1878.9	1883.9
3	1766.5	1769.7	14	1889.6	1894.1
4	1775.5	1778.4	15	1901.7	1907.0
5	1784.7	1788.1	16	1913.6	1917.6
6	1798.3	1805.2	17	1923.6	1928.4
7	1810.6	1816.4	18	1933.8	1937.4
8	1823.3	1829.9	19	1944.1	1947.7
9	1833.9	1837.2	20	1954.2	1958.2
10	1843.5	1848.1	21	1964.6	1970.6
11	1856.0	1860.1			

Solution: The computations were done by microcomputer. The program listing and the results of the run are shown below and on page 82.

```
]LIST

10  REM  SUNSPOT CYCLE
12  REM  VERSION 10/8/83/
18  REM  GET INPUT DATA
20 M = 21
30  DATA  1745, 1750.3, 1755.2, 1
     761.5, 1766.5, 1769.7, 1775.
     5, 1778.4, 1784.7, 1788.1, 1
     798.3, 1805.2
32  DATA  1810.6, 1816.4, 1823.3,
      1829.9, 1833.9, 1837.2, 184
     3.5, 1848.1, 1856.0, 1860.1,
      1867.2
34  DATA  1870.6, 1878.9, 1883.9,
      1889.6, 1894.1, 1901.7, 190
     7.0, 1913.6, 1917.6, 1923.6
36  DATA  1928.4, 1933.8, 1937.4,
      1944.1, 1947.7, 1954.2, 195
     8.2, 1964.6, 1970.6
40  DIM X(M)
50  DIM Y(M)
60  FOR J = 1 TO M
70  READ X(J),Y(J)
80  NEXT J
90  DIM R(M)
100  DIM F(M)
110  DIM A(M)
120  DIM B(M)
180  GOSUB 1000
300  REM  COMPUTE YEARLY RISE TIM
     E,FALL TIME, MIN-TO-MIN PERI
     OD, MAX-TO-MAX PERIOD
305  FOR J = 1 TO M - 1
310  LET R(J) = Y(J) - X(J):R(J) =
     R(J) + 0.0001
312 R% = 1000 * R(J):R(J) = R% /
     1000
320  LET F(J) = X(J + 1) - Y(J):F
     (J) = F(J) + 0.0001
322 F% = 1000 * F(J):F(J) = F% /
     1000
330  LET A(J) = X(J + 1) - X(J):A
     (J) = A(J) + 0.0001
332 A% = 1000 * A(J):A(J) = A% /
     1000
340  LET B(J) = Y(J + 1) - Y(J):B
     (J) = B(J) + 0.0001
342 B% = 1000 * B(J):B(J) = B% /
     1000
345  NEXT J
350  GOSUB 2000
400  REM  COMPUTE AND PRINT MEANS
410  PRINT : PRINT "MEAN";
415  DIM Z(M)
416  FOR J = 1 TO M
420  LET Z(J) = R(J)
425  NEXT J
430  GOSUB 3000
440  LET RAV = ZAV
450  HTAB 9: PRINT RAV;
455  FOR J = 1 TO M
460 Z(J) = F(J)
465  NEXT J
470  GOSUB 3000
480 FAV = ZAV
490  HTAB 16: PRINT FAV;
495  FOR J = 1 TO M
500 Z(J) = A(J)
505  NEXT J
510  GOSUB 3000
520 AAV = ZAV
```

```
530  HTAB 24: PRINT AAV;
536  FOR J = 1 TO M
540 Z(J) = B(J)
545  NEXT J
550  GOSUB 3000
560 BAV = ZAV
570  HTAB 34: PRINT BAV
600  REM  COMPUTE AND PRINT STAND
     ARD DEVIATIONS
605  DIM D(M)
610  PRINT "S.D.";
620  FOR J = 1 TO M
630 Z(J) = R(J)
640  NEXT J
650 ZAV = RAV
660  GOSUB 4000
670 RSD = ZSD
680  HTAB 9: PRINT RSD;
690  FOR J = 1 TO M:Z(J) = F(J): NEXT
     J
700 ZAV = FAV
710  GOSUB 4000
720 FSD = ZSD: HTAB 16: PRINT FSD
     ;
730  FOR J = 1 TO M:Z(J) = A(J): NEXT
     J
740 ZAV = AAV
750  GOSUB 4000
760 ASD = ZSD: HTAB 24: PRINT ASD
     ;
770  FOR J = 1 TO M:Z(J) = B(J): NEXT
     J
780 ZAV = BAV
790  GOSUB 4000
800 BSD = ZSD: HTAB 34: PRINT BSD
810  END
1000  REM  ECHO INPUT DATA
1010  HOME
1020  PRINT "CYCLE    YR OF MIN
        YR OF MAX"
1030  FOR J = 1 TO M
1040  HTAB 3: PRINT J;
1050  HTAB 12: PRINT X(J);
1060  HTAB 25: PRINT Y(J)
1070  NEXT J
1080  RETURN
2000  REM  PRINT YEARLY RISE TIME
      ,FALLTIME,MIN-MIN PERIOD,MAX
      -MAX PERIOD
2010  PRINT "ALL TIMES ARE IN YEA
      RS": PRINT
2020  PRINT "CYCLE   RISE   FALL
       MIN-MIN   MAX-MAX"
2030  HTAB 9: PRINT "TIME    TIME
       PERIOD    PERIOD"
2040  FOR J = 1 TO M - 1
2050  HTAB 3: PRINT J;
2060  HTAB 9: PRINT R(J);
2070  HTAB 16: PRINT F(J);
2080  HTAB 24: PRINT A(J);
2090  HTAB 34: PRINT B(J)
2100  NEXT J
2110  RETURN
3000  REM  COMPUTE MEAN TO TWO DE
      CIMAL PLACES
3010 SUM = Z(1)
3020  FOR J = 2 TO (M - 1)
3030 SUM = SUM + Z(J)
3040  NEXT J
3050 Z = SUM / (M - 1)
3060 Z% = 100 * Z
3070 ZAV = Z% / 100
3080  RETURN
4000  REM  COMPUTER STANDARD DEVI
      ATIONS TO 2 DECIMAL PLACES
4020 D(1) = Z(1) - ZAV
4030 SUM = D(1) * D(1)
4040  FOR J = 2 TO (M - 1)
4050 D(J) = Z(J) - ZAV
4060 SUM = SUM + (D(J) * D(J))
4070  NEXT J
4080 SD =  SQR (SUM / (M - 2))
4090 SD% = 100 * SD
4100 ZSD = SD% / 100
4110  RETURN
```

```
]RUN
CYCLE     YR OF MIN      YR OF MAX
  1         1745           1750.3
  2         1755.2         1761.5
  3         1766.5         1769.7
  4         1775.5         1778.4
  5         1784.7         1788.1
  6         1798.3         1805.2
  7         1810.6         1816.4
  8         1823.3         1829.9
  9         1833.9         1837.2
  10        1843.5         1848.1
  11        1856           1860.1
  12        1867.2         1870.6
  13        1878.9         1883.9
  14        1889.6         1894.1
  15        1901.7         1907
  16        1913.6         1917.6
  17        1923.6         1928.4
  18        1933.8         1937.4
  19        1944.1         1947.7
  20        1954.2         1958.2
  21        1964.6         1970.6
ALL TIMES ARE IN YEARS

CYCLE  RISE  FALL  MIN-MIN  MAX-MAX
       TIME  TIME  PERIOD   PERIOD
  1    5.3   4.9    10.2     11.2
  2    6.3   5      11.3     8.2
  3    3.2   5.8    9        8.7
  4    2.9   6.3    9.2      9.7
  5    3.4   10.2   13.6     17.1
  6    6.9   5.4    12.3     11.2
  7    5.8   6.9    12.7     13.5
  8    6.6   4      10.6     7.3
  9    3.3   6.3    9.6      10.9
  10   4.6   7.9    12.5     12
  11   4.1   7.1    11.2     10.5
  12   3.4   8.3    11.7     13.3
  13   5     5.7    10.7     10.2
  14   4.5   7.6    12.1     12.9
  15   5.3   6.6    11.9     10.6
  16   4     6      10       10.8
  17   4.8   5.4    10.2     9
  18   3.6   6.7    10.3     10.3
  19   3.6   6.5    10.1     10.5
  20   4     6.4    10.4     12.4

MEAN   4.52  6.45   10.97    11.01
S.D.   1.19  1.36   1.24     2.17
```

PROBLEM 8. Among the studies arising from *Landsat* observations are several concerning the evaluation of properties of snowpacks. In many areas of the world, water resources are heavily dependent on winter accumulations of snow.

Computer models are being developed whereby potential water resources can be predicted from satellite measurements of microwave emission in snow-covered areas. Predictions from such models are tested and the models refined by making comparisons with ground-based measurements of snow depth and temperature. Such measurements, when graphed, inevitably show a large amount of scatter, and it is the regression line for the data that is used as the standard for comparison.

Fig. 5.6 shows such a comparison, where the horizontal scale is temperature in degrees Kelvin. (The Kelvin scale of temperature is obtained from the Celsius scale by adding a constant, 273.15, so that 0°C = 273.15°K, and 10°C = 283.15°K.)

The data points of Fig. 5.6 are listed below. Find the parameters of the equation of the regression line.

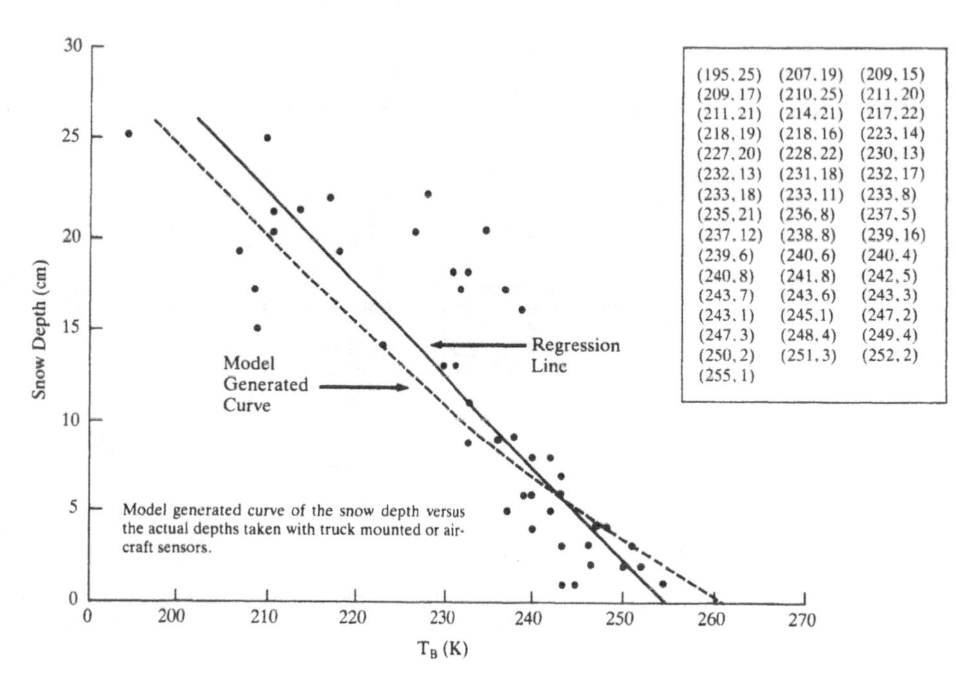

(195, 25)	(207, 19)	(209, 15)
(209, 17)	(210, 25)	(211, 20)
(211, 21)	(214, 21)	(217, 22)
(218, 19)	(218, 16)	(223, 14)
(227, 20)	(228, 22)	(230, 13)
(232, 13)	(231, 18)	(232, 17)
(233, 18)	(233, 11)	(233, 8)
(235, 21)	(236, 8)	(237, 5)
(237, 12)	(238, 8)	(239, 16)
(239, 6)	(240, 6)	(240, 4)
(240, 8)	(241, 8)	(242, 5)
(243, 7)	(243, 6)	(243, 3)
(243, 1)	(245, 1)	(247, 2)
(247, 3)	(248, 4)	(249, 4)
(250, 2)	(251, 3)	(252, 2)
(255, 1)		

Fig. 5.6

Solution: As is the previous problem, the computations were done by microcomputer. The program listing and the results follow.

```
]LIST

10  REM  SNOWPACK MODEL
15  PRINT
20  REM  REGRESSION LINE
30  DATA  195,25,207,19,209,15,20
     9,17,210,25
40  DATA  211,20,211,21,214,21,21
     7,22,218,19
50  DATA  218,16,223,14,227,20,22
     8,22,230,13
60  DATA  232,13,231,18,232,17,23
     3,18,233,11,233,8,235,21,236
     ,8,237,5,237,12
70  DATA  238,8,239,16,239,6,240,
     6,240,4,240,8,241,8,242,5,24
     3,7,243,6,243,3
80  DATA  243,1,245,1,247,2,247,3
     ,248,4,249,4,250,2,251,3,252
     ,2,255,1
90  DIM X(46): DIM Y(46)
100  FOR 1 = 1 TO 46
110  READ X(I): READ Y(I)
120  NEXT I
130  GOSUB 1000
140 XSUM = 0:YSUM = 0:SP = 0:SQ =
     0
150  FOR I = 1 TO 46
160 XSUM = XSUM + X(I):YSUM = YSU
     M + Y(I)
170 SP = SP + X(I) * Y(I):SQ = SQ
      + X(I) * X(I)
180  NEXT I
190 XMEAN = XSUM / 46:YMEAN = YSU
     M / 46
200 B = (SP - 46 * XMEAN * YMEAN)
      / (SQ - 46 * XMEAN * XMEAN)

210 B% = 1000 * B:B = B% / 1000
220 A% = 100 * YMEAN:A = A% / 100

225 C% = 100 * XMEAN:C = C% / 100

230  PRINT " THE REGRESSION LINE
     HAS SLOPE ";B
240  PRINT " AND A MEAN SNOW DEPT
     H OF ";A
250  PRINT " CORRESPONDS TO A MEA
     N TEMPERATURE": PRINT " OF "
     ;C
260  END

1000  REM  ECHO INPUT DATA
1010  PRINT "  J","X(I)" "Y(I)"
1020  FOR I = 1 TO 46
1030  PRINT I,X(I),Y(I)
1040  NEXT I
1050  PRINT "END OF DATA"
1060  RETURN
```

```
]RUN

  I                X(I)             Y(I)
1                  195              25
2                  207              19
3                  209              15
4                  209              17
5                  210              25
6                  211              20
7                  211              21
8                  214              21
9                  217              22
10                 218              19
11                 218              16
12                 223              14
13                 227              20
14                 228              22
15                 230              13
16                 232              13
17                 231              18
18                 232              17
19                 233              18
20                 233              11
21                 233              8
22                 235              21
23                 236              8
24                 237              5
25                 237              12
26                 238              8
27                 239              16
28                 239              6
29                 240              6
30                 240              4
31                 240              8
32                 241              8
33                 242              5
34                 243              7
35                 243              6
36                 243              3
37                 243              1
38                 245              1
39                 247              2
40                 247              3
41                 248              4
42                 249              4
43                 250              2
44                 251              3
45                 252              2
46                 255              1
END OF DATA
 THE REGRESSION LINE HAS SLOPE -.45
 AND A MEAN SNOW DEPTH OF 11.3
 CORRESPONDS TO A MEAN TEMPERATURE
 OF 232.63
```

CHAPTER SIX

EXPONENTIAL AND LOGARITHMIC FUNCTIONS

APOLLO 17 EVA—Astronaut Eugene A. Cernan, Commander of the mission is photographed by Astronaut Schmitt whose photo is reflected in the gold visor.

The early work that led to our understanding of the planetary motions and gave us the description of the solar system we know today would have been virtually impossible without the use of logarithms to reduce the labor of the computations. Although computers and calculators have replaced logarithms as computational tools, logarithmic and exponential functions are still essential for the study of Earth's atmosphere and rocket propulsion, examples of which are cited in this chapter.

PROBLEM 1. Experimentation and theory have shown that an approximate rule for atmospheric pressure at altitudes less than 80 km is the following: Standard atmospheric pressure, 1035 grams per square centimeter, is halved for each 5.8 km of vertical ascent.

a. Write a simple exponential equation to express this rule.

Solution: Letting P denote atmospheric pressure at altitudes less than 80 km and h the altitude in km, we have

$$P = 1035\,(1/2)^{h/5.8}\ \text{g/cm}^2.$$

b. Compute the atmospheric pressure at an altitude of 40 km.

Solution: From the equation of part (a),

$$\begin{aligned} P &= 1035\ (1/2)^{40/5.8}\ \text{g/cm}^2 \\ &= 1035\ (1/2)^{6.9}\ \text{g/cm}^2 \\ &= 1035\ (0.0084)\ \text{g/cm}^2 \\ &= 8.7\ \text{g/cm}^2. \end{aligned}$$

c. Find the altitude at which the pressure is 20 percent of standard atmospheric pressure.

Solution: Substituting in the equation of part (a) gives $(0.20)(1035) = (1035)(1/2)^{h/5.8}$, where h is in km, and so $(0.2) = (1/2)^{h/5.8}$. Now, taking logarithms,

$$\log(0.2) = \frac{h}{5.8}\log(0.5)$$

and

$$h = 5.8\frac{\log(0.2)}{\log(0.5)}\ \text{km} = 5.8\,(2.32)\ \text{km} = 13.5\ \text{km}.$$

PROBLEM 2. The rule for the variation of atmospheric pressure with height which was given in the previous problem can also be written

$$\begin{aligned} P &= 1035\,(2)^{-h/5.8} \\ &= 1035\,(2)^{-0.17h}. \end{aligned}$$

Atmospheric scientists often use this rule in one of its equivalent forms where the base is 10 or e, the base of the natural logarithms, instead of 2. Find k_1 and k_2 so that $P = 1035\,(2)^{-0.17h} = 1035\,(10)^{-k_1h} = 1035\,(e)^{-k_2h}$.

Solution: We need to find k_1 so that $2^{0.17} = 10^{k_1}$. Taking logarithms, $0.17 \log 2 = k_1$ or $k_1 = (0.17)(0.301) = 0.051$. For k_2 we have $2^{0.17} = e^{k_2}$, or $k_2 = (0.17)\log_e 2 = (0.17)(0.693) = 0.12$.

PROBLEM 3. Sometimes different bases are used together in the same application in atmospheric work. For example, atmospheric absorption of electromagnetic radiation from the Sun and other sources is dependent on the wavelengths of the incoming radiation. Instruments carried by rockets, balloons, and satellites have shown how far in the atmosphere such radiation penetrates before being reduced by a factor of $1/e$, the conventional measure used in this work. The results are given in Fig. 6.1. Both the wavelength scale and the altitude scale are logarithmic, with the horizontal scale in base 10 and the vertical scale in base 2. (How much of this information could be displayed using linear scales even on a wall-sized chart?)

Fig. 6.1 shows that visible light and radio waves penetrate the atmosphere completely and reach Earth's surface. However, gases such as oxygen, ozone, nitrogen, and water vapor absorb most of the infrared, ultraviolet, X-ray, and shorter wavelengths. At what altitude will solar infrared radiation of wavelength 10^{-4} m be reduced by a factor of $1/e$?

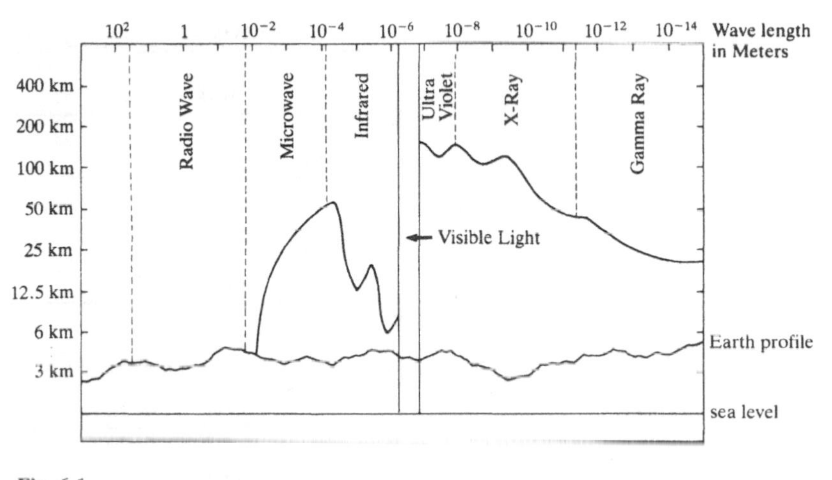

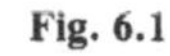

Fig. 6.1

Solution: The equal intervals on the altitude scale have length log 2. The ordinate we are seeking, y, is 1/4 of the way between log 50 and log 100. This means that

$$\log y = \log 50 + \frac{1}{4}\log 2 = 1.699 + \frac{1}{4}(0.3010)$$

$$= 1.699 + 0.075 = 1.774.$$

$$\text{Then } y = 10^{1.774} = 59.$$

If a calculator with a y^x key is available, we can solve this problem without actually finding logarithms, as follows:

$$\log y = \log 50 + \frac{1}{4}\log 2 = \log\left[(50)(2)^{1/4}\right] = \log[(50)(1.189)]$$

$$= \log 59. \qquad \text{So } y = 59.$$

In the foregoing problem, we saw how the use of logarithmic scales made it possible to display information over an extremely large range of values. The next two problems show another use for logarithmic scales, that of fitting a mathematical function to experimental data.

PROBLEM 4. Very high energy particles (electrons and protons) are found in the radiation belts of some planets (e.g., Earth, Jupiter, Saturn), and a plot of the number of particles found at different energies is called a *spectrum.* Often the spectrum has a shape that can be represented by an equation of the form $N = KE^m$ where N is the number of particles at a certain energy, E; K is a proportionality factor; and m is called the spectral index.

When the spectrum has such a shape, we call it a power-law spectrum, and the experimenter studying such a spectrum wants to know the values of m and K. Table 6.1 shows values of N measured at several Es during the flight of *Pioneer 10* past Jupiter. For these data, find the best value of m and of K. (N is really the number of particles hitting a detector per unit time, or the counting rate, which is why the number can be a fraction.)

Table 6.1

Energy, E	Number, N
0.16	1.0×10^6
0.30	1.5×10^5
0.60	1.3×10^4
1.0	6.8×10^3
1.6	1.0×10^3
4.5	20
10.0	1
20.0	0.1

Solution: Using logarithms on the expression $N = KE^m$ results in $\log N = \log K + m \log E$, or, to obtain the form of a linear equation $y = mx + b$, $\log N = m \log E + \log K$.

We can find logarithms for the values of E and N in the table (or we can use log-log graph paper and circumvent this step), plot the points, and draw the best straight line through this set of points. Then m will be the slope of the line, and K will be the value of N for which $\log E = 0$. (Note that this is the value that lies on the best straight line, and not necessarily any value in the data set.)

We observe in Fig. 6.2 that the intercept on the $\log N$ scale is 3.5. Since $\log N = 3.5$ when $\log E = 0$, we have $\log K = 3.5$, so $K = 10^{3.5} = 3200$. The points $(0, 3.5)$ and $(1.0, 0)$ are on the best fit line, so $m = \dfrac{3.5 - 0}{0 - 1.0} = -3.5$. So $N = (3200)\, E^{-3.5}$

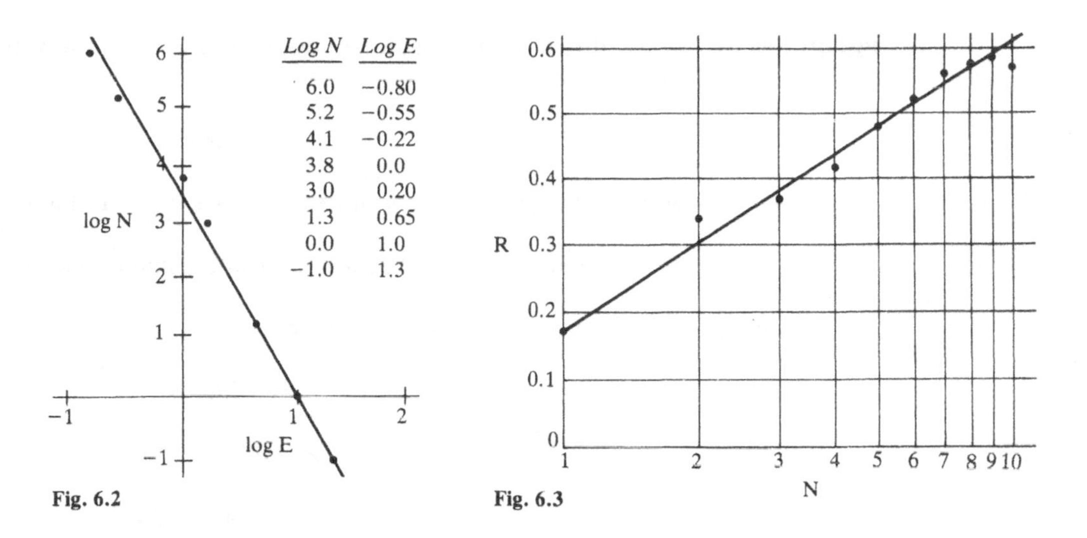

Fig. 6.2

Fig. 6.3

Many of the control functions in a space vehicle system are automatic, handled by computers and electronic feedback devices. However, the adaptability and the decision-making ability of human monitors of these systems, whether crew members or ground controllers, greatly increase the overall effectiveness of mission control. Because of this, managers of projects in the space program have become interested in some results from psychological studies of human decision making.

PROBLEM 5. One such study measured the time it took to respond when faced with varying numbers of choices. Experimental results are given in the table below, where N is the number of choices presented and R is the reaction time in seconds. Graph these data on semilogarithmic graph paper with N on the logarithmic scale (or graph R against $\log_{10}N$ if semilog graph paper is not available) and find an empirical expression for reaction time as a function of the number of choices.

N	1	2	3	4	5	6	7	8	9	10
R	0.17	0.34	0.37	0.42	0.48	0.52	0.56	0.58	0.59	0.57

Solution: The points are graphed and a "best fit" line drawn (see Fig. 6.3). Since the point $(N, R) = (1, 0.17)$ does lie on this line, we have $R = 0.17 + m \log_{10}N$. To find m, we can use the points $(1, 0.17)$ and $(9, 0.59)$, since both are on the "best fit" line:

$$0.59 = 0.17 + m \log_{10}9$$

so

$$0.42 = m\,(0.954),$$

or

$$m = (0.42)/(0.954) = 0.44$$

The requested relation is $R = 0.17 + 0.44 \log_{10}N$.

As we have seen in Chapter 4, solar cells, which convert solar energy into electrical energy, can be used to supply power in space vehicles. Nuclear energy derived from radioactive isotopes is also used. Nuclear energy sources gradually lose power in a manner described by the exponential function. The next problem illustrates some computations of the available power and operational life of a satellite using a nuclear power source.

PROBLEM 6. A satellite has a radioisotope power supply. The power output in watts is given by the equation

$$P = 50e^{-t/250}$$

where t is the time in days and e is the base of natural logarithms.

a. How much power will be available at the end of one year?

Solution: Applying the given equation, we have

$$\begin{aligned} P &= 50e^{-365/250} \\ &= 50e^{-1.46} \\ &= 50 \times 0.232 \\ &= 11.6 \end{aligned}$$

Thus approximately 11.6 watts will be available at the end of one year.

b. What is the half-life of the power supply? In other words, how long will it take for the power to drop to half its original strength?

Solution: To find the half-life, we solve the equation

$$25 = 50e^{-t/250}$$

for t and obtain

$$\begin{aligned} \frac{-t}{250} &= \ln 0.5 \\ &= -0.693 \\ t &= 250 \times 0.693 \\ &= 173. \end{aligned}$$

Thus the half-life of the power supply is approximately 173 days. (Note that $\ln x$ is a shorter expression for $\log_e x$.)

c. The equipment aboard the satellite requires 10 watts of power to operate properly. What is the operational life of the satellite?

Solution: Solving the equation

$$10 = 50e^{-t/250}$$

for t gives

$$\frac{-t}{250} = \ln\frac{10}{50}$$

$$= \ln 0.2$$

$$= -1.609$$

$$t = 250 \times 1.609$$

$$= 402.$$

Hence the operational life of the satellite is 402 days.

PROBLEM 7. The theory of rocket flight shows that the velocity gained by a launch vehicle when its propellant is burned to depletion is expressed by the equation

$$v = c \ln R$$

where v is the velocity gained by the vehicle during launch;
c is the exhaust velocity of the engine;
$\ln R$ is $\log_e R$, or the natural logarithm of R;

and R is the mass ratio of the spacecraft, defined by $R = \dfrac{\text{takeoff weight}}{\text{burnout weight}}$.

a. The takeoff weight consists of propellant or fuel, F, structure, S, and payload, P. At burnout, assuming all the fuel has been used, the remaining weight is $S + P$, so that $R = \dfrac{F + S + P}{S + P}$. In general, the weight of fuel cannot be more than about 10 times the weight of the structure in order for the vehicle to withstand the stresses of operation. Show that if $F = 10S$, then an upper limit for R is 11.

Solution: If $F = 10S$, then $R = \dfrac{F + S + P}{S + P} = \dfrac{10S + S + P}{S + P}$

$$= \frac{11(S + P) - 10P}{S + P}$$

$$= 11 - \frac{10P}{S + P} \leq 11.$$

So the largest possible value for R is 11, but we see that in order to actually achieve this value, it is necessary for P to be 0—in other words, the launch vehicle could carry no payload!

b. The minimum altitude for a stable orbit about Earth is about 160 km. At lower altitudes, air resistance slows the spacecraft and causes a rapid deterioration of the orbit. As will be shown in Problem 1 of Chapter 9, the spacecraft must attain a velocity of about 7.8 km per second to orbit at 160 km. However, in order to overcome the retarding effect of Earth's atmosphere while the spacecraft is ascending, the total velocity imparted by the launch vehicle must be at least 9.0 km/s. What is the minimum exhaust velocity needed by the rocket engine if $R = 11$?

Solution: Substituting

$$v = 9.0 \text{ km/s}$$

and $R = 11$ in the rocket equation,

$$9.0 \text{ km/s} = c \ln 11$$

$$c = \frac{9.0}{\ln 11} \text{ km/s} = \frac{9.0}{2.4} = 3.8 \text{ km/s}.$$

c. The propellants used for engines such as those of the Delta, Centaur, and Saturn launch vehicles could produce exhaust velocities averaging at most 3 km/s, which would not be sufficient to achieve orbit. The main engines of the Space Shuttle use a mixture of liquid hydrogen and liquid oxygen, which can produce exhaust velocities of 4.6 km/s. However, in order for the Shuttle to perform its tasks and return to Earth with its crew, it has an R-value of around 3.5. Could the Space Shuttle achieve orbit with its main engines?

Solution: If

$$c = 4.6 \text{ km/s, and}$$

$$R = 3.5, \text{ then}$$

$$\begin{aligned} v &= 4.6 \ln 3.5 \text{ km/s} \\ &= (4.6)(1.25) \text{ km/s} \\ &= 5.8 \text{ km/s,} \end{aligned}$$

which is not sufficient for orbit.

PROBLEM 8. It is apparent from the rocket equation that the burnout velocity increases when the mass ratio increases. We can get a higher mass ratio by using a solid propellant because the stiff, rubberlike propellant mass serves as part of the structure. If no payload, or a very small payload, is included, a solid-propellant rocket could have a mass ratio of about 19. A typical average exhaust velocity for a solid propellant might be about 2.4 km per second. Could this launch vehicle achieve a 160 km Earth orbit?

Solution: Using the rocket equation,

$$v = 2.5 \ln 19 \text{ km/s}$$

$$= (2.4)(2.94)$$

$$= 7.1 \text{ km/s},$$

which is much less than that needed for orbit.

The solution to the problem pointed out in the preceding examples is to use staging. That is, the launch vehicle is divided into two or more parts, or stages. As soon as the propellant has been burned in the first stage, there is a brief coast during which the heavy motors and structure in the first stage are jettisoned and permitted to fall into the ocean. Freed from this deadweight, the second-stage motors are much more effective; the same procedure is repeated for the remaining stages.

PROBLEM 9. Let us design a two-stage vehicle to place a payload into Earth orbit. We shall make some simplifying assumptions to make this problem easier while preserving the basic idea: (1) the structure weight of each stage is 10 percent of the fuel weight, the remaining weight being payload, (2) the gain in velocity is divided equally among the stages, each contributing 4.5 km/s to the required final velocity of 9.0 km/s; (3) all stages use the same propellant with an exhaust velocity of 3.7 km/s. This third assumption is generally not true in practice—for example, the Space Shuttle uses solid rocket boosters in addition to the main engines—but our goal here is to see how staging works. For the sake of having a numerical example, we shall also assume that the total weight at liftoff is 5.0×10^4 kg. For this numerical example, determine the weight of fuel to be carried by each stage, the structural weight of each stage, and the weight of the orbital payload.

Solution: Let F_1, S_1, P_1 represent fuel, structure, and payload weight, respectively, of the first stage, and F_2, S_2, and P_2 those of the second stage. Since the "payload" of the first stage includes the entire second stage and the orbital payload,

$$P_1 = F_2 + S_2 + P_2.$$

First stage:

$$v = c \ln R_1$$

$$4.5 = 3.7 \ln R_1$$

$$\ln R_1 = \frac{4.5}{3.7} = 1.22$$

$$R_1 = e^{1.22} = 3.4$$

So

$$\frac{F_1 + S_1 + P_1}{S_1 + P_1} = \frac{5.0 \times 10^4}{S_1 + P_1} = 3.4$$

Then

$$S_1 + P_1 = \frac{5.0 \times 10^4}{3.4} = 1.5 \times 10^4 \text{ kg}$$

and

$$F_1 = (5.0 - 1.5) \times 10^4 = 3.5 \times 10^4 \text{ kg}.$$

By assumption 1,

$$S_1 = 0.10\,(3.5 \times 10^4) = 3.5 \times 10^3 \text{ kg}.$$

Then

$$P_1 = 1.5 \times 10^4 - 3.5 \times 10^3 = 1.15 \times 10^4 \text{ kg}.$$

Second stage: We again have, from the rocket equation,

$$4.5 = 3.7 \ln R_2,$$

so

$$R_2 = 3.4.$$

Also,

$$R_2 = \frac{1.15 \times 10^4}{S_2 + P_2}.$$

Then

$$S_2 + P_2 = \frac{1.15 \times 10^4}{3.4} = 3.4 \times 10^3 \text{ kg}.$$

Therefore,

$$F_2 = 1.15 \times 10^4 - 3.4 \times 10^3 = 8.1 \times 10^3 \text{ kg}$$

$$S_2 = 0.10\,(8.1 \times 10^3) = 0.8 \times 10^3 \text{ kg}$$

$$P_2 = (3.4 - 0.8) \times 10^3 \text{ kg} = 2.6 \times 10^3 \text{ kg}.$$

Our design for the two-stage launch vehicle may be checked as follows:

Weight of fuel:	kg × 10^3
F_1	35.0
F_2	8.1
Total	43.1
Weight of structure:	
S_1	3.5
S_2	0.8
Total	4.3
Weight of orbital payload	2.6
Total weight of vehicle	50.0 = 5.0 × 10^4 kg

Thus, although the single-stage launch vehicle discussed in Problem 7 could not place any payload into orbit, this two-stage vehicle can place nearly 5 percent of its weight into Earth orbit.

PROBLEM 10. Show that when all stages use the same propellant, the total mass ratio of a multiple-stage launch vehicle is equal to the product of the individual mass ratios.

Solution: Indicate the burnout velocities and mass ratios of the first, second, third stages, and so on, by the subscripts 1, 2, 3, and so on. Then, using a three-stage vehicle as an example,

$$v_1 + v_2 + v_3 = c \ln R_1 + c \ln R_2 + c \ln R_3$$

$$v = c\,(\ln R_1 + \ln R_2 + \ln R_3)$$

$$v = c \log_e(R_1 R_2 R_3)$$

(*Note:* Making the structure stronger so that it can support large payloads reduces the mass ratios. However, if we have several stages, the total mass ratio can become very high, producing much greater performance.)

PROBLEM 11. Using the equation derived in Problem 9, show that the launch vehicle constructed in Problem 8 can indeed orbit its payload.

Solution: Given

$$R_1 R_2 = (3.4)(3.4) = 11.56$$

$$v = 2.7 \log_e 11.56$$

$$= 3.7(2.45)$$

$$= 9.06 \text{ km/s}$$

The launch vehicle will impart sufficient velocity to overcome drag losses and insert the payload into a 160-km Earth orbit. Note that dividing the launch vehicle into stages increases the overall mass ratio to 11.56.

CHAPTER SEVEN

TRIGONOMETRY

Close-up view of the 64-meter tracking antenna of the Deep Space Network located at Goldstone, California.

Angle measurements and the trigonometric analysis of such measurements are used extensively in space science. Among the examples we shall consider here are some involving transformations between terrestrial (or celestial) and spacecraft coordinate systems, a variety of photogrammetric corrections, and the tracking of spacecraft from stations on Earth.

PROBLEM 1. A conventional right-handed three-dimensional spacecraft coordinate system is shown in Fig. 7.1. The angular motions of the spacecraft with respect to the x-, y-, and z-axes respectively are called *roll*, *pitch*, and *yaw*, shown in Fig. 7.1 by curved arrows. We shall develop the transformations between this coordinate system in a moving spacecraft and a reference coordinate system whose origin coincides with the one in the diagram but does not undergo rotation. Here, we shall consider a single rotation at a time. In Chapter 8, "Matrix Algebra," we shall investigate a series of such rotations.

When the spacecraft performs a rotation, the reference system remains fixed, but the spacecraft coordinate system undergoes the same rotation as the spacecraft. If the point Q has coordinates (x, y, z) in the reference system, we need to find its coordinates in the spacecraft system after such a rotation takes place. Let us consider each of the motions roll, pitch, and yaw separately.

a. Let the spacecraft coordinate system initially coincide with the reference system, and let the spacecraft undergo roll through angle R. Express the coordinates (x_R, y_R, z_R) of a point Q on the spacecraft in terms of (x, y, z) and R after this motion is performed.

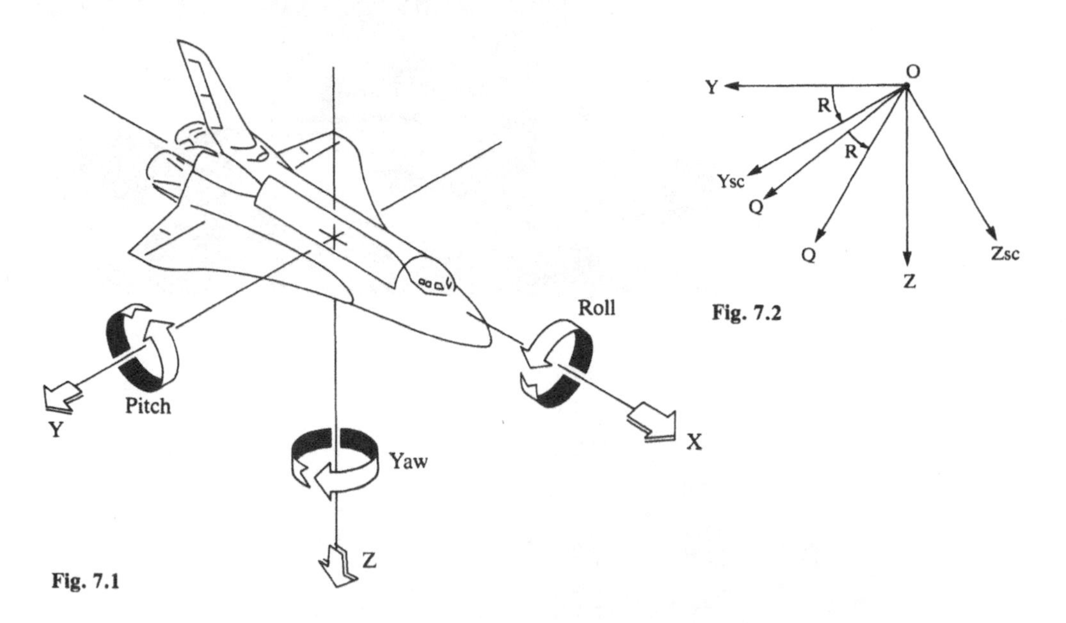

Fig. 7.1

Fig. 7.2

Solution: Since the roll is around the x-axis, the x-coordinate of Q is the same in both systems: $x_R = x$. Now consider the plane parallel to the y-z plane, which contains Q. The roll moves Q to Q′ as shown in Fig. 7.2. Let $r = OQ = OQ'$ and let $\angle YOQ' = \theta$. Then $\angle Y_{sc}OQ' = \theta - R$. Q′ has coordinates (y, z) in the reference system, where

$$y = r\cos\theta \quad \text{and} \quad z = r\sin\theta.$$

In the spacecraft system, Q′ has coordinates (y_R, z_R)

where $\quad y_R = r \cos(\theta - R) \quad$ and $\quad z_R = r \sin(\theta - R)$.

Expanding the sine and cosine of this difference results in

$$y_R = r \cos\theta \cos R + r \sin\theta \sin R = y \cos R + z \sin R$$

$$z_R = r \sin\theta \cos R - r \cos\theta \sin R = z \cos R - y \sin R.$$

b. Find the comparable transformations if the rotation is either a pitch through an angle P or a yaw through an angle Y.

Solution: For a *pitch* rotation, this takes place around the y-axis, so if the coordinates in the spacecraft system are (x_P, y_P, z_P), we have $y_P = y$. We next consider a plane parallel to the x-z plane, and the analysis will be just as in part (a) with y replaced by z, z replaced by x, and $\angle R$ replaced by $\angle P$, resulting in $x_P = x \cos P - z \sin P$; $z_P = z \cos P + x \sin P$.

A *yaw* rotation takes place around the z-axis, so if the coordinates in the spacecraft system are (x_Y, y_Y, z_Y), we have $z_Y = z$; now we consider a plane parallel to the x-y plane, and this time the analysis is just as in part (a), with y replaced by x, z replaced by y, and $\angle R$ replaced by $\angle Y$. The result is $x_Y = x \cos Y + y \sin Y$; $y_Y = y \cos Y - x \sin Y$. (We note that the right-handed system dictates that a positive angle of rotation take place so that the cyclic order x y z x is maintained.)

c. An Earth-based computer monitoring the coordinates of Jupiter in *Voyager*'s reference frame recorded Jupiter at (2.03, −2.81, 0.336) (in units equivalent to 10^5 km) at one point. If *Voyager* had performed a yaw rotation of 28° just prior to this reading, what were Jupiter's coordinates in the spacecraft coordinate system?

Solution: Using (x', y', z') for the spacecraft coordinate system:

$$\begin{aligned} x' &= x \cos 28° + y \sin 28° \\ &= 2.03 \cos 28° - 2.81 \sin 28° \\ &= 0.473 \\ y' &= -x \sin 28° + y \cos 28° \\ &= -2.03 \sin 28° - 2.81 \cos 28° \\ &= -3.43 \\ z' &= z = 0.336 \end{aligned}$$

So the coordinates were (0.473, −3.43, 0.336) in these units.

We next calculate the length of some of the latitude circles on Earth.

PROBLEM 2. Although Earth is not really a sphere, it can be treated as though it were spherical for many purposes.

a. Show that the length of any parallel of latitude around Earth is equal to the equatorial distance around Earth times the cosine of the latitude angle (see Fig. 7.3), if we assume a spherical shape for Earth.

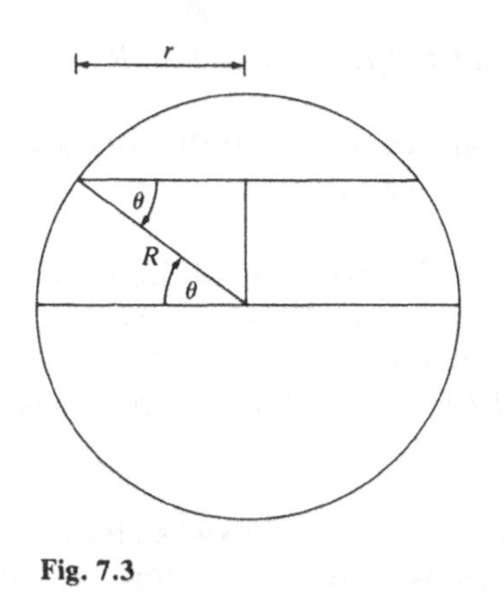

Fig. 7.3

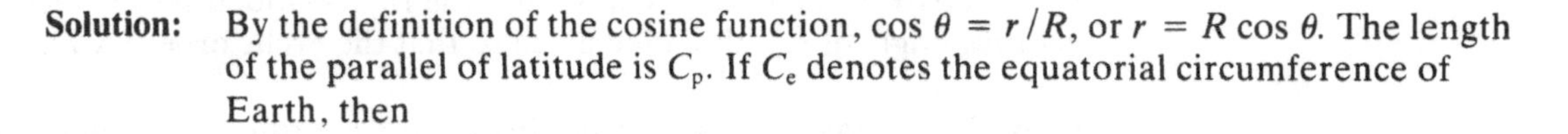

Solution: By the definition of the cosine function, $\cos\theta = r/R$, or $r = R\cos\theta$. The length of the parallel of latitude is C_p. If C_e denotes the equatorial circumference of Earth, then

$$C_p = 2\pi r$$

$$= 2\pi R\cos\theta$$

$$= C_e\cos\theta.$$

b. Find the length of the 30° parallel, north or south latitude. Use $R = 6400$ km.

Solution: Applying the formula for the length of a parallel of latitude derived in part (a) gives

$$C_p = (6400\text{ km})(\cos 30°)$$

$$= (6400\text{ km})(0.866)$$

$$= 5500\text{ km}.$$

c. Determine the length of the Arctic Circle (66°33′ N).

Solution: Using the formula from part (a), the length is

$$C_p = (6400\text{ km})(\cos 66°33')$$

$$= (6400\text{ km})(0.398)$$

$$= 2500\text{ km}.$$

d. How far is it "around the world" along the parallel of 80° north latitude?

Solution: Using the result of part (a), the distance is

$$C_p = (6400 \text{ km})(\cos 80°)$$

$$= (6400 \text{ km})(0.1737)$$

$$= 1100 \text{ km}.$$

PROBLEM 3. Two tracking stations s miles apart measure the elevation angle of a weather balloon to be α and β, respectively (Fig. 7.4). Derive a formula for the altitude h of the balloon in terms of the angles α and β. Ignore Earth's curvature.

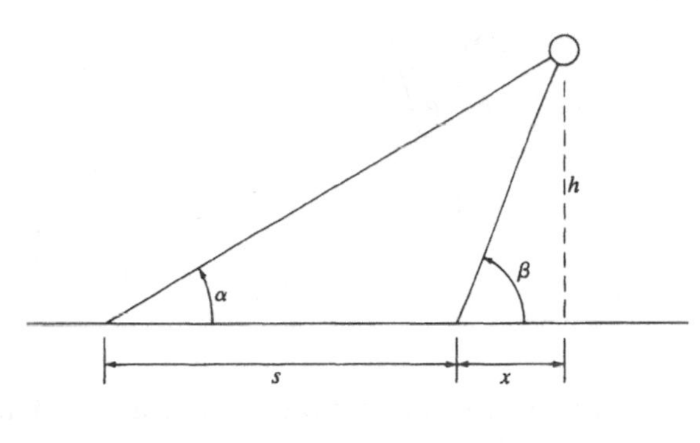

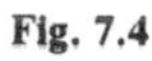
Fig. 7.4

Solution: Writing an equation for the cotangent of each angle and solving for x gives

$$\cot \alpha = \frac{s + x}{h}$$

$$x = h \cot \alpha - s$$

and

$$\cot \beta = \frac{x}{h}$$

$$x = h \cot \beta.$$

Now the two expressions for x are equated:

$$h \cot \alpha - s = h \cot \beta$$

so

$$h(\cot \alpha - \cot \beta) = s$$

and

$$h = \frac{s}{\cot \alpha - \cot \beta}$$

PROBLEM 4. A satellite traveling in a circular orbit 1600 km above Earth is due to pass directly over a tracking station at noon. Assume that the satellite takes two hours to make an orbit and that the radius of Earth is 6400 km.

a. If the tracking antenna is aimed 30° above the horizon, at what time will the satellite pass through the beam of the antenna? (See Fig. 7.5.)

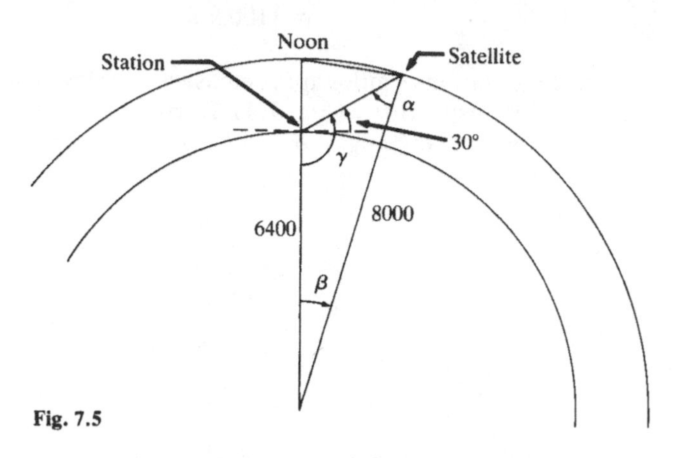

Fig. 7.5

Solution: In the triangle formed by the station, the satellite, and the center of Earth, $\gamma = 120°$. From the law of sines,

$$\frac{\sin \alpha}{6400} = \frac{\sin \gamma}{8000}$$

$$\sin \alpha = \frac{6400 \sin 120°}{8000} = 0.693.$$

Then

$$\alpha = 44°$$

and

$$\beta = 180° - (120° + 44°) = 16°.$$

The time between

$$\beta = 16° \text{ and } \beta = 0.0° \text{ is } \frac{16°}{360°}(120 \text{ min})$$

$$= 5.3 \text{ min}.$$

This means that the satellite will pass through the beam of the antenna at 12:00 − 5.3 minutes, or 11:54.7 a.m.

b. Find the distance between the satellite and the tracking station at 12:03 p.m.

Solution: Computing angle β gives

$$\beta = \frac{3 \text{ min}}{120 \text{ min}} 360° = 9°.$$

By the law of cosines,

$$x^2 = (6400)^2 + (8000)^2 - 2(6400)(8000) \cos 9°$$

$$= (40.96 + 64 - 101.14) \times 10^6 \text{ km}^2$$

$$= 3.82 \times 10^6 \text{ km}^2$$

$$x = 1.96 \times 10^3 = 2.0 \times 10^3 \text{ km}.$$

We have found that the distance between the satellite and the tracking station is 2000 km (to two significant figures) at 12:03 p.m.

c. At what angle above the horizon should the antenna be pointed so that its beam will intercept the satellite at 12:03 p.m.? (See Fig. 7.6.)

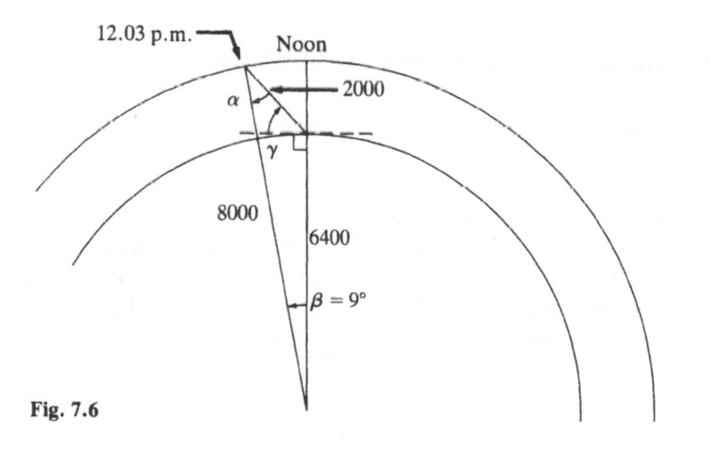

Fig. 7.6

Solution: Again, applying the law of sines,

$$\frac{\sin 9°}{2000} = \frac{\sin(\gamma + 90°)}{8000}$$

$$\sin(\gamma + 90°) = \frac{8000}{2000} \sin 9° = 0.626$$

$$\cos \gamma = 0.626$$

$$\gamma = 51°.$$

PROBLEM 5. Two of NASA's tracking stations are located near the equator; one is in Ethiopia, at 40° east longitude, another near Quito, Ecuador, at 78° west longitude. Assume both stations, represented by E and Q in Fig. 7.7, are on the equator and that the radius of Earth is 6380 km. A satellite in orbit over the equator is observed at the same instant from both tracking stations. The angles of elevation above the horizon are 5° from Quito and 10° from Ethiopia. Find the distance of the satellite from Earth at the instant of observation.

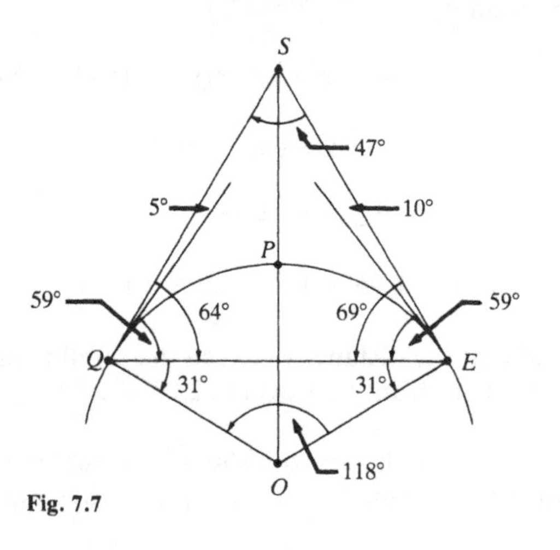

Fig. 7.7

Solution: In Fig. 7.7, OQ = OP = OE = 6380 km: ∠QOE = the longitude difference of the two stations, so ∠QOE = 78° − (−40°) = 118°. Since ΔQEO is isosceles,

$$\angle OQE = \angle OEQ = \frac{1}{2}(180° - 118°) = 31°.$$

Further, since the horizon is perpendicular to the radius, ∠SQE = 5° + (complement of ∠EQO) = 5° + 59° = 64°, and ∠SEQ = 10° + (complement of ∠QEO) = 10° + 59° = 69°. Also, ∠QSE = 180° − (64° + 69°) = 47°. These angles are all shown in Fig. 7.7. We are looking for the distance SP. If we can determine OS, then SP = OS − OP = OS − 6380 km. We note that OS is not an angle bisector for either ∠QOE or ∠QSE, so we must use an indirect method to find OS.

We can evaluate QE from $\dfrac{QE}{\sin 118°} = \dfrac{OE}{\sin 31°}$:

$$QE = \frac{6380 \sin 118°}{\sin 31°} = 1.094 \times 10^4 \text{ km};$$

then

SE from $\dfrac{SE}{\sin 64°} = \dfrac{QE}{\sin 47°}$:

$$SE = \frac{1.094 \times 10^4 \sin 64°}{\sin 47°} = 1.34 \times 10^4 \text{ km};$$

now

$$OS = \sqrt{(OE)^2 + (SE)^2 - 2(OE)(SE)\cos\angle OES}$$

$$= \sqrt{(6.38 \times 10^3)^2 + (1.34 \times 10^4)^2 - (2)(6.38 \times 10^3)(1.34 \times 10^4)\cos 100^\circ}$$

$$= 10^3\sqrt{40.70 + 119.68 + 24.24}$$

$$= 10^3\sqrt{249.62} = 1.58 \times 10^4\,\text{km}.$$

So

$$SP = 15800 - 6400 = 9400\,\text{km}.$$

Although the Sun is more than a hundred times as large as Earth, as we noted in the first problem of Chapter 4, it subtends an angle of only about half a degree in the sky as viewed from Earth. In the next problem, we consider some aspects of the observation of sunspots.

PROBLEM 6. **a.** Find the angular separation between two large sunspots when viewed from Earth (or Earth orbit) if they are separated by 30° in longitude along the Sun's equator. Consider two cases:

1. A time when the midpoint between the spots is on the center of the visible disc of the Sun;

2. A time about a week later when the Sun has rotated so that the leading spot is just about to go over the Sun's limb (edge).

Recall that the Earth-Sun distance is 1.5×10^8 km. The radius of the Sun is 7.0×10^5 km. In the first case it will help, and in the second it will be necessary, to make a suitable approximation (Fig. 7.8).

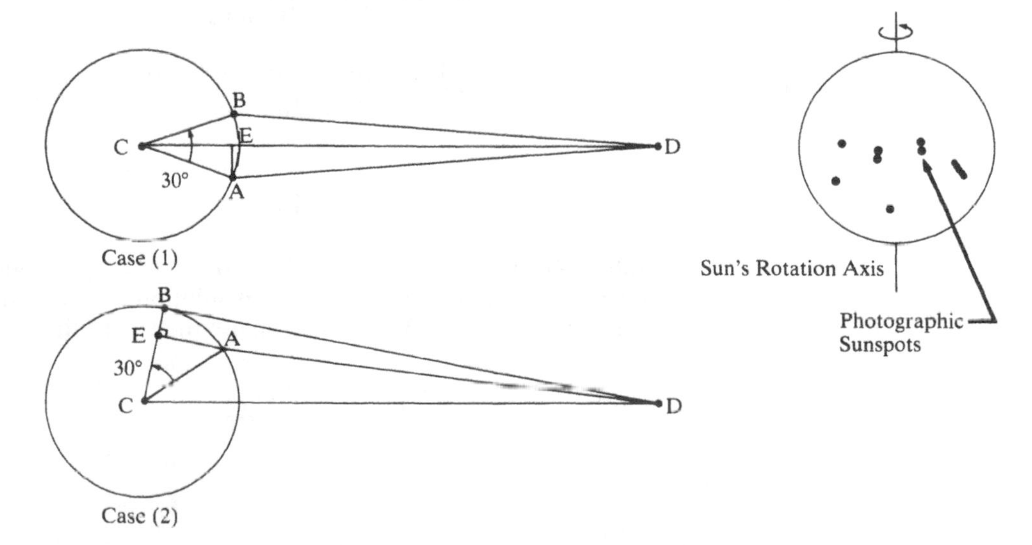

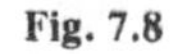

Fig. 7.8

Solution: **Case 1.** In the edge-on drawing shown above we have:

$$CD = \text{Earth-Sun distance} = 1.5 \times 10^8 \text{ km}$$

$$CA = CB = \text{radius of Sun} = 7.0 \times 10^5 \text{ km}$$

$$\angle ABC = 30^\circ \text{ and CD bisects } \angle ACB$$

Let AE be the perpendicular from A to CD and let h be its length.

Then

$$CE = h \cot \angle ACE;$$

$$ED = h \cot \angle ADE;$$

and

$$h = CA \sin \angle ACE.$$

So

$$\begin{aligned} CD &= CE + ED \\ &= h \cot \angle ACE + h \cot \angle ADE \\ &= CA \sin \angle ACE \cot \angle ACE + CA \sin \angle ACE \cot \angle ADE. \end{aligned}$$

Then

$$\begin{aligned} \cot \angle ADE &= \frac{CD - CA \sin \angle ACE \cot \angle ACE}{CA \sin \angle ACE} = \frac{CD - CA \cos \angle ACE}{CA \sin \angle ACE} \\ &= \frac{1.5 \times 10^8 - 7.0 \times 10^5 \cos 15^\circ}{7.0 \times 10^5 \sin 15^\circ} \\ &= \frac{1.5 \times 10^8 - 7.0 \times 10^5 \times 0.97}{7.0 \times 10^5 \times 0.26} \\ &= \frac{1.5 \times 10^8 - 6.8 \times 10^5}{1.8 \times 10^5} = \frac{1.5 \times 10^8}{1.8 \times 10^5} = 0.82 \times 10^3. \end{aligned}$$

$\angle ADE = \text{arccot } 0.82 \times 10^3 = 0.070^\circ$, so the angular separation between the sunspots $= 2 \angle ADE = 0.14^\circ$. A simpler solution can be found if we approximate AD by saying $AD \doteq CD$. Now we can use the law of sines:

$$\frac{CA}{\sin \angle ADC} = \frac{AD}{\sin \angle ACD},$$

then

$$\sin \angle ADC = \frac{CA}{AD} \sin \angle ACD = \frac{CA}{CD} \sin \angle ACD$$

$$= \frac{7.0 \times 10^5}{1.5 \times 10^8} \sin 15° = 4.7 \times 10^{-3} \times 0.26$$

$$= 1.2 \times 10^{-3};$$

and so

$$\angle ADC = 0.07° \text{ and } \angle ADB = 2 \angle ADC = 0.14°.$$

Case 2. There is more than one way to solve this, but we present just one solution and use an approximation. In the drawing for Case 2 (see Fig. 7.8), construct the perpendicular AE from A to BC. For the approximation, we shall use $\tan \angle ADB = \frac{EB}{BD}$. In ΔAEC, $\cos \angle ACE = \frac{CE}{CA}$, so $CE = CA \cos \angle ACE = 7.0 \times 10^5 \cos 30° = 7.0 \times 10^5 \times 0.87 = 6.1 \times 10^5$ km.

Then $EB = CB - CE = 7.0 \times 10^5 - 6.1 \times 10^5 = 0.9 \times 10^5 = 9 \times 10^4$ km.

Now, from our approximation, $\tan \angle ADB = \frac{9 \times 10^4}{1.5 \times 10^8} = 6 \times 10^{-4}$, giving the angular separation $\angle ADB = 0.036°$.

b. The unaided eye can distinguish a sunspot if it is 1.5 minutes of arc, or 0.025 degrees, across. Sunspot sizes are usually measured in units of 0.001 of the Sun's area. What is the minimum size of sunspot that can be seen without a telescope?

Solution: Sun's area $= 4\pi r^2 = 4\pi(7.0 \times 10^5)^2$ km^2 $= 196\pi \times 10^{10}$ km^2. Since 1 sunspot unit $= 10^{-3}$ of the Sun's area, we have 1 sunspot unit $= 196\pi \times 10^7$ km^2. Now if we assume that we have a sunspot that is approximately a disc subtending an angle of 0.025° at Earth, we see from Fig. 7.9 that the disc has radius = (Earth-Sun distance) $\times \sin\left(\frac{0.025°}{2}\right)$

$$= 1.5 \times 10^8 \times 2.2 \times 10^{-4} \text{ km} = 3.3 \times 10^4 \text{ km}.$$

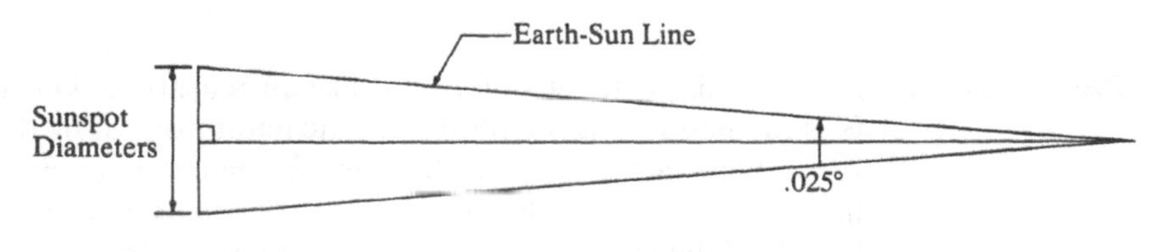

Fig. 7.9

Such a disc has area $\pi(3.3 \times 10^4)^2$ km^2

$$= \pi(3.3)^2 \times 10^8 \text{ km}^2 \times \frac{1 \text{ sunspot unit}}{\pi \times 196 \times 10^7 \text{ km}^2}$$

$$= \frac{108.9 \times 10^7}{196 \times 10^7} \text{ sunspot units}$$

$$= 0.55 \text{ sunspot units.}$$

Historical note: Very few sunspots exceed an angular diameter of 1.5 minutes of arc. Normally, the Sun is too dazzling to permit an observation of such a sunspot by the unaided eye; however, if the Sun is low on the horizon and shines through a thick haze, sunspots can be observed. Pretelescopic sunspot observations have been recorded by Chinese and Japanese viewers. (**Caution:** Never look directly at the Sun.)

The photographic scale factor for vertical aerial photographs was developed in Problem 7 of Chapter 4. We now consider the situation when the camera is tilted so that the film is not parallel to the ground. The result of such tilting is shown in Fig. 7.10, where the broken lines represent a square grid as it would appear in a vertical photograph and the solid lines show the actual image on a tilted photograph. (This is sometimes called the "keystone effect.") In order to use the photograph to produce an undistorted picture, numerical relationships must be established between the actual shapes and their photographic images.

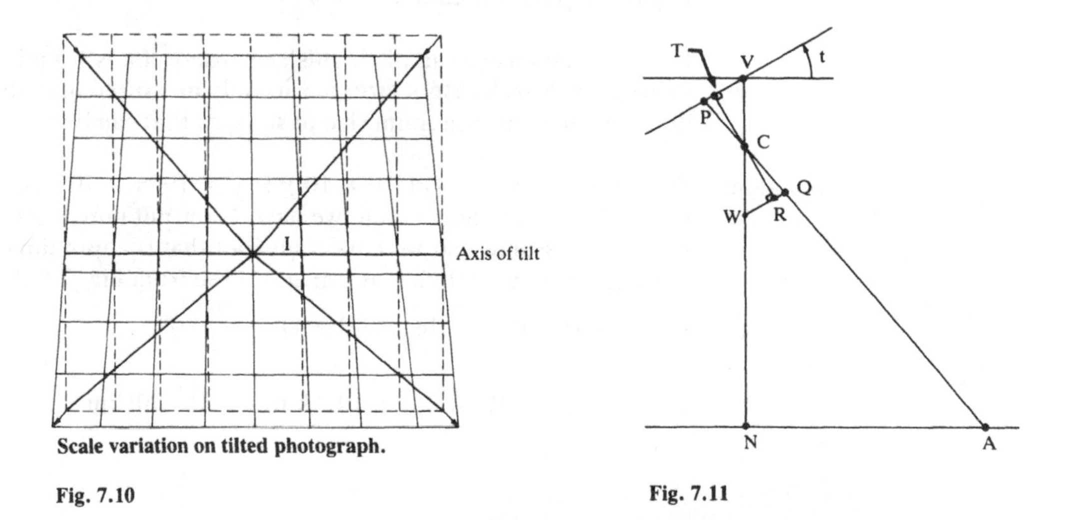

Scale variation on tilted photograph.

Fig. 7.10

Fig. 7.11

PROBLEM 7. Fig. 7.11 shows the geometry of the configuration, where the camera is located at C, N is the nadir, V is the photographic nadir point, P is the image of ground point A, and t is the tilt angle of the film (the acute angle made by the film with the horizontal). If CT is the normal from the camera to the film, CT = f, the focal length of the camera. CN = H is the height of the camera above the ground, which we assume to be level. It is customary in this work to use the film "positive" QW instead of the "negative" PV. This is obtained by choosing W on CN and Q on CA so that CW = CV and CQ = CP. We let R be the point on the film positive so that CR is normal to the film.

a. Let θ be the angle made by the line from the camera to the ground point A with respect to the vertical, where $\theta > 0$ if Q and R are on the same side of CN and $\theta < 0$ if they are on opposite sides of CN ($\theta = \angle$NCQ). Express the ratio of the length of the image WQ to the length of NA in terms of θ, t, f, and H for the case where $\theta > t$.

Solution: From the geometry, we see that since CR $\perp$ WQ, $\angle$NCR = $\angle$VCT = t and

$$\frac{\text{QW}}{\text{AN}} = \frac{\text{QR} + \text{RW}}{\text{AN}} = \frac{f \tan(\theta - t) + f \tan t}{H \tan \theta}.$$

$$= \frac{\dfrac{f(\tan\theta - \tan t)}{1 + \tan\theta \tan t} + f \tan t}{H \tan\theta}$$

$$= \frac{f \tan\theta - f \tan t + f \tan t + f \tan\theta \tan^2 t}{H \tan\theta (1 + \tan\theta \tan t)}$$

$$= \frac{f \tan\theta (1 + \tan^2 t)}{H \tan\theta (1 + \tan\theta \tan t)}$$

$$\frac{\text{QW}}{\text{AN}} = \frac{f(1 + \tan^2 t)}{H(1 + \tan\theta \tan t)}.$$

b. Show that if $t = 0$ (untilted camera) or $t = \theta$ (camera aimed at point A), then $\frac{\text{QW}}{\text{AN}} = \frac{f}{H}$. (Recall from Problem 7 of Chapter 4 that this is the scale factor of a vertical photograph.)

Solution: For $t = 0$, $\tan t = 0$ and the result follows. For $t = \theta$, the second factor in the denominator becomes $(1 + \tan^2 t)$, which cancels, and the result follows.

c. Show that the result of part (a) still applies for the cases where Q is between W and R (Fig. 7.12) and where Q is on the side of CN that does not contain R (Fig. 7.13), taking into consideration the sign of θ.

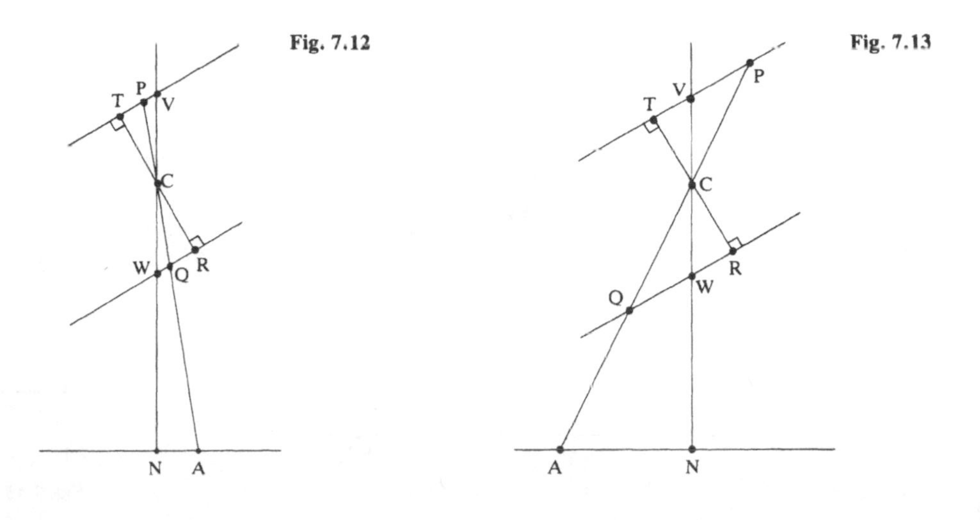

Fig. 7.12

Fig. 7.13

Solution: If Q is between W and R, then

$$\frac{QW}{AN} = \frac{RW - RQ}{AN} = \frac{f \tan t - f \tan (t - \theta)}{H \tan \theta}$$

$$= \frac{f\,(1 + \tan^2 t)}{H\,(1 + \tan \theta \tan t)}.$$

If Q is as shown in Fig. 7.13, then θ is negative, and the positive value of the angle in the diagram is $(-\theta)$.

So

$$\frac{QW}{AN} = \frac{QR - RW}{AN} = \frac{f \tan (\,(-\theta) + t\,) - f \tan t}{H \tan (-\theta)}$$

$$\frac{f \tan (t - \theta) - f \tan t}{-H \tan \theta} = \frac{f\,(1 + \tan^2 t)}{H\,(1 + \tan t \ \tan \theta)}.$$

Note that this implies (since θ is negative in the last case) that points on the "down" side of the film will have their images "stretched out," whereas points on the "up" side (at least those for which $\theta > t$) will have their images "shrunk."

d. Fig. 7.10 showed a point I (called the isocenter) at which there is no distortion in the scale of the tilted photograph. Show that I is the point of intersection of the bisector of ∠NCR in Fig. 7.11 with the film positive QW by establishing that a vertical photograph taken with the camera in its position at C would contain the point I.

Solution: The bisector of ∠NCR is shown in Fig. 7.14, along with a horizontal through I that intersects CN at Y. Since CN is vertical and IY is horizontal, ∠CYI is a right angle. Triangles CYI and CRI have corresponding angles equal and share side CI and are therefore congruent. Since CY = CR = f, a vertical photographic positive and the actual photograph positive from the same camera position C both contain the point I.

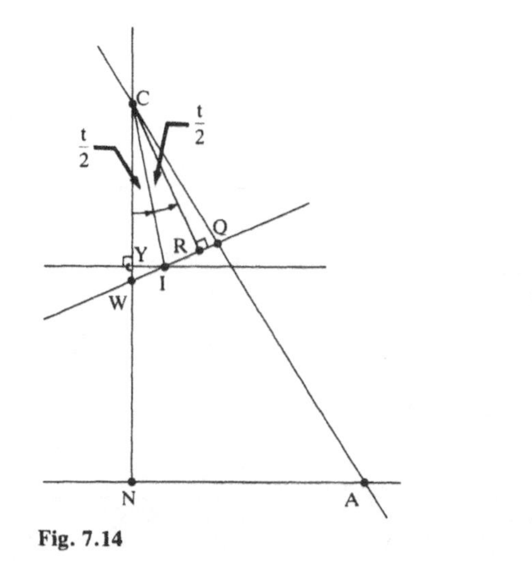

Fig. 7.14

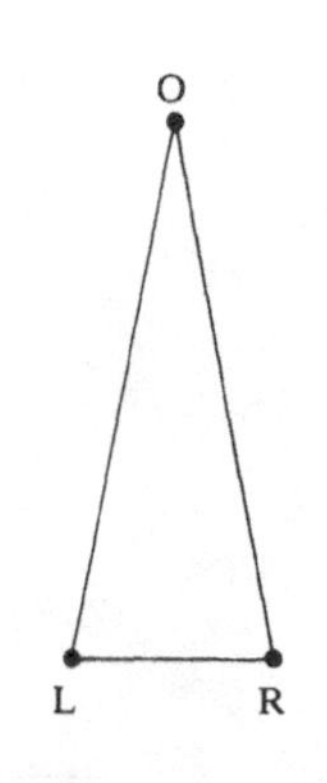

Fig. 7.15

In Problem 8 of Chapter 4 we developed a formula to correct an aerial photograph for distortion due to Earth's curvature. The distortion occurs because the camera cannot distinguish how far away an object is—it has no "depth perception." For aerial photography, the picture is interpreted as though everything is in the plane tangent to Earth at the nadir; in satellite photography, as we shall see in Problem 9, pictures will be interpreted (unless corrected) as though everything is in the horizon plane sensed by the satellite.

PROBLEM 8. Depth perception in humans has two aspects, called *monoscopic* and *stereoscopic*. Monoscopic judgments of distance use only one eye and are based on an interpretation of relative sizes of objects, shadows, hidden portions of objects, and other attributes of this type; such judgments are very rough and frequently fail. Stereoscopic judgments of distance use both eyes and are quite accurate in most people. Stereoscopic judgment depends on the physical separation of the eyes, which causes an object to be viewed at a different angle by each eye, as shown in Fig. 7.15. The angle subtended by the "eye base" (the distance LR where L is the left eye and R the right) at the object O is called the *parallactic angle*; it is evident that the closer the object, the larger the parallactic angle.

The smallest parallactic angle discernible by human eyes is about 0.025°, and the average adult eyes are spaced about 6.5 cm apart. What is the largest distance at which the average adult can judge depth?

Solution: Let d be the distance of O from LR in Fig. 7.15. We present two methods of solution. The first uses the fact that

$$\tan\left(\frac{1}{2}\angle \mathrm{LOR}\right) = \left(\frac{1}{2}\,\mathrm{LR}\right)\Big/ d, \text{ so } d = (0.0325)/\tan(0.0125°)\text{ m}$$

$$\doteq 150\text{ m}.$$

For another approach, we may approximate LR as an arc of a circle with radius d where LR subtends an angle $\theta = 0.025°$. If θ is in radians, then $\mathrm{LR} = \theta \cdot d$. So $\theta = 0.025° = \frac{0.025}{180} = 0.00044$ rad and therefore

$$d = \frac{0.025}{0.00044}\text{ m} \doteq 150\text{ m}.$$

Satellites such as the *Landsats*, *Seasat*, and the *Synchronous Meteorological Satellites* (SMS-1 and -2) have made it possible to study Earth and its oceans, resources, and weather patterns as never before. They have returned observations and data that are being used by botanists, geologists, oceanographers, and meteorologists, among others, in numerous projects. To cite just two examples, *Landsat* observations have been used in the assessment of soil moisture in agricultural fields, and SMS observations have been useful in predicting severe storms.

PROBLEM 9. A spacecraft at a distance h from Earth in synchronous orbit can see only a portion of Earth's surface, as illustrated in Fig. 7.16. The circle that is the boundary of this spherical "cap" will be called the *horizon circle*, and the spacecraft has sensors that can recognize this horizon.

Although every spacecraft uses its horizon sensors to find its angular direction with respect to Earth's center, those satellites whose purpose is to observe Earth can also use this angle measurement to determine the size of the spherical cap that can be observed.

In Fig. 7.17, S is the position of the spacecraft, C is the center of Earth, H is a point on the horizon circle seen by the spacecraft, P is the subsatellite point on Earth (the intersection of Earth's surface with the line from Earth's center to the satellite), and Q is the center of the horizon circle. We have SP = h and CH = CP = r, the radius of Earth. ρ is the angular separation of the horizon seen by the spacecraft from Earth's center, and λ is the angle subtended at Earth's center by the radius of the horizon circle.

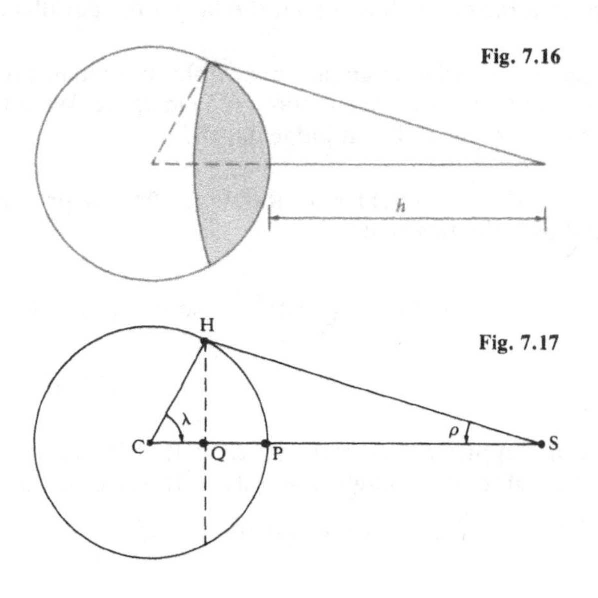

Fig. 7.16

Fig. 7.17

a. Find the relationships among ρ, λ, h, and r.

Solution: Since ΔSHC has a right angle at H, $\sin \rho = \cos \lambda = \dfrac{r}{r + h}$.

b. Listed below are some Earth-observing satellites and their perigee or apogee distances from Earth. For each, find the angular radius (λ) of the horizon circle seen by the spacecraft. (Earth's radius is 6378 km.)

Landsat 2	916 km (apogee)	*OGO-1*	260 km (perigee)
Seasat	790 km (apogee)	*OGO-1*	150 000 km (apogee)
SMS-2	36 000 km (apogee)		

(OGO is the *Orbiting Geophysical Observatory*.)

Solution: Using $\lambda = \cos^{-1}\left(\frac{r}{r+h}\right)$, for *Landsat 2* we have

$$= \cos^{-1}\left(\frac{6378}{6378 + 916}\right) = \cos^{-1}(0.8744) = 29°.$$

Similarly, we get angular radii of 27°, 81°, 16°, and 87° for the remaining cases, respectively.

c. If a satellite sees a horizon circle of angular radius 30°, what is its distance from Earth?

Solution:

$$\cos 30° = \frac{6378}{6378 + h} = 0.8660$$

$$h = \frac{6378(1 - 0.8660)}{0.8660} = 987 \text{ km (to the nearest km).}$$

In observing Earth from space using spacecraft sensors, distortions are introduced because of Earth's spherical shape. For example, suppose a thick black line is painted along the equator, the 10° parallel of latitude, and the 50° and 90° west meridians of longitude as shown in Fig. 7.18(a). Uncorrected observations of this "rectangle" would appear as shown in Fig. 7.18(b). The diagram in Fig. 7.19 illustrates how this distortion comes about. Although spacecraft sensors can measure the angle at which point R on Earth is observed, they cannot measure the distance to R—all observations are interpreted as though lying in the same plane, so the image of R is treated as though it were at R′, in the plane of the horizon circle.

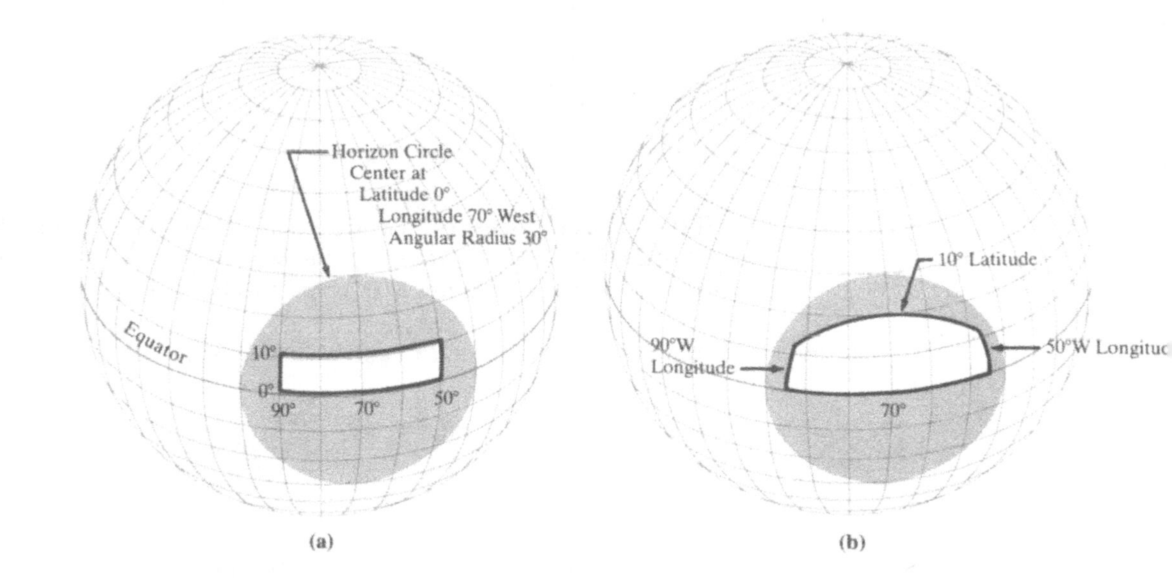

Fig. 7.18

The data can be corrected by the spacecraft's computers so that the information relayed to Earth is distortion-free. The actual computer program that does the correction depends also on the particular hardware of the sensors, but the first step in the correction is to express the relationship among the angle of observation of R ($\angle\alpha$), the angular deviation of R from the line joining Earth's center to the satellite ($\angle\beta$), and the angle of observation of the horizon ($\angle\rho$). Since α and ρ can be measured, the computer can then find β for the proper mapping of R.

d. Show that the relationship linking α, β, and ρ is given by

$$\tan\alpha = \frac{\sin\rho\sin\beta}{1 - \sin\rho\cos\beta}.$$

Solution: In Fig. 7.19, if T is the foot of the perpendicular from R to CS, then

$$\tan\alpha = \frac{\text{RT}}{\text{TS}} = \frac{\text{RT}}{\text{CS} - \text{CT}} = \frac{r\sin\beta}{(r+h) - r\cos\beta}$$

$$= \frac{\frac{r}{r+h}\sin\beta}{1 - \frac{r}{r+h}\cos\beta} = \frac{\sin\rho\sin\beta}{1 - \sin\rho\cos\beta}$$

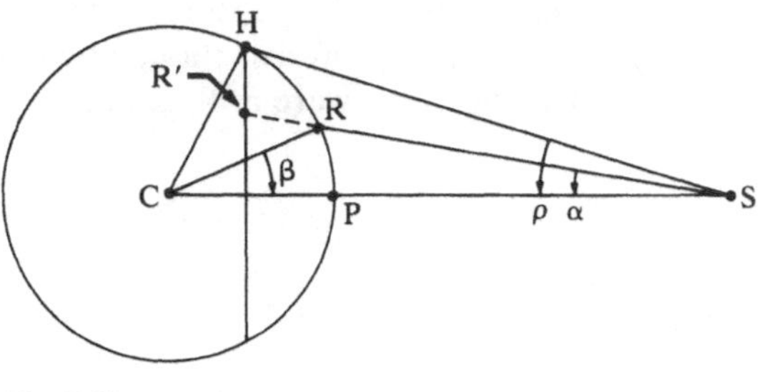

Fig. 7.19

e. If the spacecraft sensors measure $\angle\rho$ as 30° and a point R is observed at an angle of 25° from the subsatellite point, what is the actual angular displacement of R from the subsatellite point with respect to Earth's center?

Solution: We have $\angle\rho = 30°$, $\angle\alpha = 25°$, and we are seeking $\angle\beta$.

From the last equation, $\tan 25° = \dfrac{\sin 30°\sin\beta}{1 - \sin 30°\cos\beta}$,

so

$$0.446 = \frac{(0.5)\sin\beta}{1 - (0.5)\cos\beta}$$

$$= \frac{\sin\beta}{2 - \cos\beta};$$

then

$$(0.466)^2 = \frac{\sin^2 \beta}{4 - 4\cos\beta + \cos^2\beta},$$

$$0.217 = \frac{1 - \cos^2\beta}{4 - 4\cos\beta + \cos^2\beta}.$$

Clearing fractions gives $0.868 - 0.868\cos\beta + 0.217\cos^2\beta = 1 - \cos^2\beta$, and collecting terms gives $1.217\cos^2\beta - 0.868\cos\beta - 0.132 = 0$.

Then

$$\cos\beta = \frac{0.868 \pm \sqrt{(-0.868)^2 - 4(1.217)(-0.132)}}{2(1.217)}$$

$$= \frac{0.868 \pm \sqrt{1.396}}{2.438}.$$

Since we know that $|\beta| < 90°$, we discard the negative root, and so

$$\cos\beta = \frac{2.050}{2.438} = 0.841$$

$$\beta = 32.8° \doteq 33°.$$

We have already seen in Fig. 2.2 of Chapter 2 that the celestial coordinate system uses angles of declination and right ascension in a manner analogous to the latitude and longitude angles of the coordinate system of Earth. We now compare the three-dimensional spherical coordinate system commonly used in mathematics with the one generally used in astronomy and space science.

PROBLEM 10. Texts in analytic geometry or calculus with analytic geometry usually define a spherical coordinate system so that if for $P(\rho, \theta, \phi)$ we let Q be the foot of the perpendicular from P to the x-y plane (Fig. 7.20), then

ρ = the distance OP, $\rho \geq 0$

θ = the angle made by OQ with the positive x-axis, the positive angular direction being a rotation from OX toward OY, $0 \leq \theta < 2\pi$

ϕ = the angle made by OP with the z-axis, with the positive angular direction being away from OZ, $0 \leq \phi \leq \pi$.

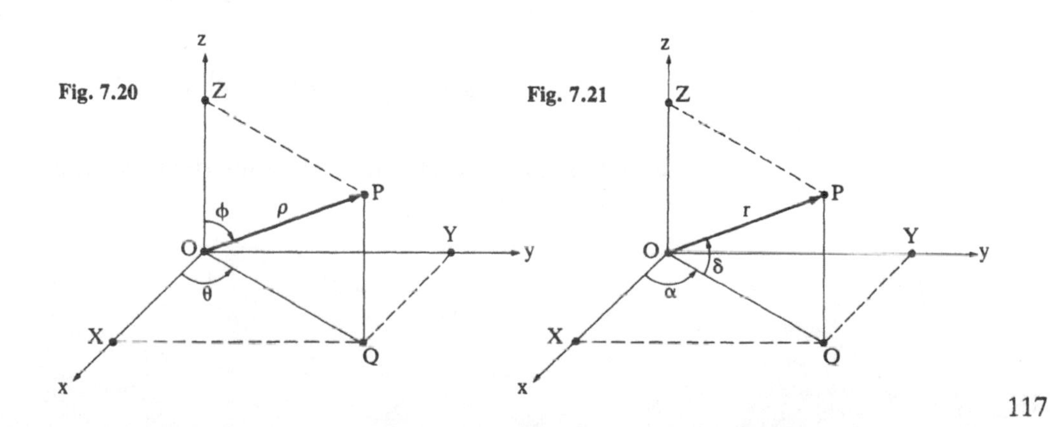

Fig. 7.20

Fig. 7.21

In this system, as the reader may verify, the transformation between (x, y, z) and (ρ, θ, ϕ) are as follows:

$$x = \rho \sin \phi \cos \theta \qquad \rho = (x^2 + y^2 + z^2)^{\frac{1}{2}}$$

$$y = \rho \sin \phi \sin \theta \qquad \theta = (\arctan (y/x)$$

$$z = \rho \cos \phi \qquad \phi = \arccos \left(z/(x^2 + y^2 + z^2)^{\frac{1}{2}} \right)$$

In the spherical coordinate system used by astronomers and space scientists, if P has coordinates (r, δ, α) and Q is the foot of the perpendicular from P to the x-y plane (Fig. 7.21), then

r = the distance OP, $r \geq 0$

δ = the angle made by OP with OQ, the positive angular direction being from OQ toward the positive z-axis,

$$-\frac{\pi}{2} \leq \delta \leq \frac{\pi}{2}$$

α = the angle made by OQ with the positive x-axis, the positive angular direction being a rotation from OX toward OY, $0 \leq \alpha < 2\pi$.

Develop the transformations from (r, δ, α) to (x, y, z).

Solution: From the definitions, it is evident that

$$\rho = r, \theta = \alpha, \phi = \frac{\pi}{2} - \delta.$$

So we have

$$x = \rho \sin \phi \cos \theta = r \sin \left(\frac{\pi}{2} - \delta \right) \cos \alpha = r \cos \delta \cos \alpha$$

$$y = \rho \sin \phi \sin \theta = r \sin \left(\frac{\pi}{2} - \delta \right) \sin \alpha = r \cos \delta \sin \alpha$$

$$z = \rho \cos \phi = r \cos \left(\frac{\pi}{2} - \delta \right) = r \sin \delta$$

(Recall that δ is the declination and α the right ascension in the celestial coordinate system.)

PROBLEM 11. On March 5, 1979, the spacecraft *Voyager 1* passed close to the Jovian moon Io. This close encounter took place just after *Voyager*'s closest radial approach to Jupiter, which occurred at about noon on that day. If we set up a Cartesian coordinate system centered at Jupiter with the x-y plane as Io's orbital plane and the Jupiter-to-Sun vector as the positive x-axis (see Fig. 7.22), then *Voyager*'s spherical coordinates at t_0 = 13 hours were $r = 5.0$, $\delta = -5.0°$, and $\alpha = 127°$. (We measure lengths in units of Jovian radii, R_J, where $1\ R_J = 70\,000$ km. The spherical coordinate system used here is the one defined in the previous problem.)

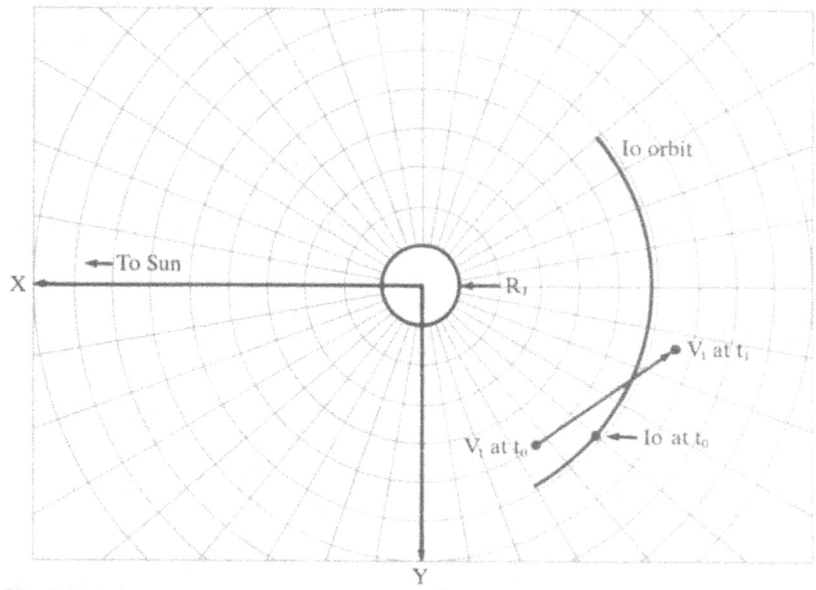

Fig. 7.22

a. What was the *Voyager*'s radial distance from Jupiter in km at t_0 = 13 hours?

Solution:

$$r = 5.0\ (R_J) = 5.0 \times 70\,000 \text{ km} = 350\,000 \text{ km}$$

b. What were its Cartesian (x, y, z) coordinates in the system defined above?

Solution:

$$x = 5.0 \cos(-5.0°) \cos(127°) = -3.0\ R_J$$

$$y = 5.0 \cos(-5.0°) \sin(127°) = 4.0\ R_J$$

$$z = 5.0 \sin(-5.0°) = -0.44\ R_J$$

Three hours later, at t_1 = 16 hours, its coordinates were

$$r = 6.5,\ \delta = -1.9°,\ \alpha = 166°.$$

c. What were *Voyager*'s Cartesian coordinates at $t_1 = 16$ hours?

Solution:

$$x = 6.5 \cos(-1.9°) \cos(166°) = -6.3\ R_J$$

$$y = 6.5 \cos(-1.9°) \sin(166°) = 1.6\ R_J$$

$$z = 6.5 \sin(-1.9°) = -0.22\ R_J$$

As you can see, in the interval *Voyager* has moved away from Jupiter in the anti-Sun direction (its x-coordinate has become more negative), toward the Sun-Jupiter line (its y-coordinate has decreased), and it has moved toward Io's orbital plane (its z-coordinate has decreased in abolute value).

If we assume that *Voyager*'s Cartesian coordinates change linearly with time between t_0 and t_1, this means that we assume that *Voyager* has constant velocity components in the x, y, and z directions.

d. Under this assumption, what are *Voyager*'s velocity components in the x, y, and z directions?

Solution:

$$V_x = \frac{x_1 - x_0}{t_1 - t_0} = \frac{-6.3 - (-3.0)}{16 - 13} = -1.1\ R_J/\text{h}$$

$$= 1.1 \times 70\,000\ \text{km/h} = 77\,000\ \text{km/h}$$

$$V_y = \frac{y_1 - y_0}{t_1 - t_0} = \frac{1.6 - (4.0)}{16 - 13} = -0.8\ R_J/\text{h} = 56\,000\ \text{km/h}$$

$$V_z = \frac{z_1 - z_0}{t_1 - t_0} = \frac{-0.22 - (0.44)}{16 - 13} = 0.073\ R_J/\text{h} = 5\,100\ \text{km/h}$$

e. Under the assumption that *Voyager*'s Cartesian coordinates vary linearly with time, find expressions for $x(t)$, $y(t)$, and $z(t)$.

Solution: Since $x = x_0 + V_x(t - t_0)$, we have $x = -3.0 - 1.1(t - t_0)$. Similarly, $y = 4.0 - 0.8(t - t_0)$ and $z = -0.44 + 0.073(t - t_0)$. While *Voyager* was moving, Io had been progressing in its orbit. Consider Io's orbit to be a circle of radius $r = 5.9$ and recall that in this coordinate system, Io's δ equals 0 at all times. At $t_0 = 13$ hours on 5 March, Io's phase angle α was 139°.

f. This phase angle is a linear function of time. Knowing that Io's orbital period is 42.5 hours (i.e., it takes Io 42.5 hours to move 360° in α), derive an expression for $\alpha(t)$.

Solution: Io moves through $\frac{360}{42.5}$ degrees per hour, and $\alpha = 139°$ at t_0, so

$$\alpha(t) = \frac{360\,(t - t_0)}{42.5} + 139\ \text{degrees} = 8.47\,(t - t_0) + 139\ \text{degrees}.$$

g. Find Io's rectangular coordinates as functions of time.

Solution:

$$x = r \cos \delta \cos \alpha = 5.9 \cos (8.47 (t - t_0) + 139), \text{ since } \cos \delta = 1$$

$$y = r \cos \delta \sin \alpha = 5.9 \sin (8.47 (t - t_0) + 139)$$

$$z = r \sin \delta = 0$$

h. Derive an expression for the separation distance Δ between *Voyager* and Io as a function of time. Use ζ for $(t - t_0)$.

Solution:

$$\Delta^2 = (x_{\text{Voy}} - x_{\text{Io}})^2 + (y_{\text{Voy}} - y_{\text{Io}})^2 + (z_{\text{Voy}} - z_{\text{Io}})^2$$

From parts (e) and (g),

$$\Delta^2 = [-3.0 - 1.1\zeta - 5.9 \cos (8.47\zeta + 139)]^2$$

$$+ [4.0 - 0.8\zeta - 5.9 \sin (8.47\zeta + 139)]^2$$

$$+ [-0.44 + 0.073\zeta]^2$$

i. Use a calculator and evaluate Δ, for several values of ζ, in the interval $0 \leq \zeta \leq 3$. Plot the results, and use the resulting graph to find when *Voyager*'s closest approach to Io occurs and at what distance.

Solution:

ζ	Δ
0	1.51
0.5	1.23
1.0	0.94
1.5	0.62
2.0	0.33
2.1	0.28
2.2	0.26
2.3	0.25
2.4	0.27
2.5	0.31
2.75	0.46
3.0	0.65

Fig. 7.23

The graph is shown in Fig. 7.23.

We see that the closest approach occurred at $t = 13 + 2.3$ hours $= 15.3$ hours on March 5 at a distance of $0.25\ R_J$, or about 17 500 km.

j. What are the components of the *Voyager*-Io separation vector at the time of closest approach?

Solution:

$$\Delta_x = -3.0 - 1.1(2.3) - 5.9 \cos (8.47 \times 2.3 + 139)^\circ = -0.041\ R_J$$

$$\Delta_y = 4.0 - 0.8(2.3) - 5.9 \sin (8.47 \times 2.3 + 139)^\circ = +0.004\ R_J$$

$$\Delta_z = -0.44 + 0.073(2.3) = -0.27\ R_J$$

Thus, *Voyager 1* was mostly "below" Io at closest approach: its separation was almost entirely in the *z* direction, perpendicular to Io's orbital plane.

CHAPTER EIGHT

MATRIX ALGEBRA

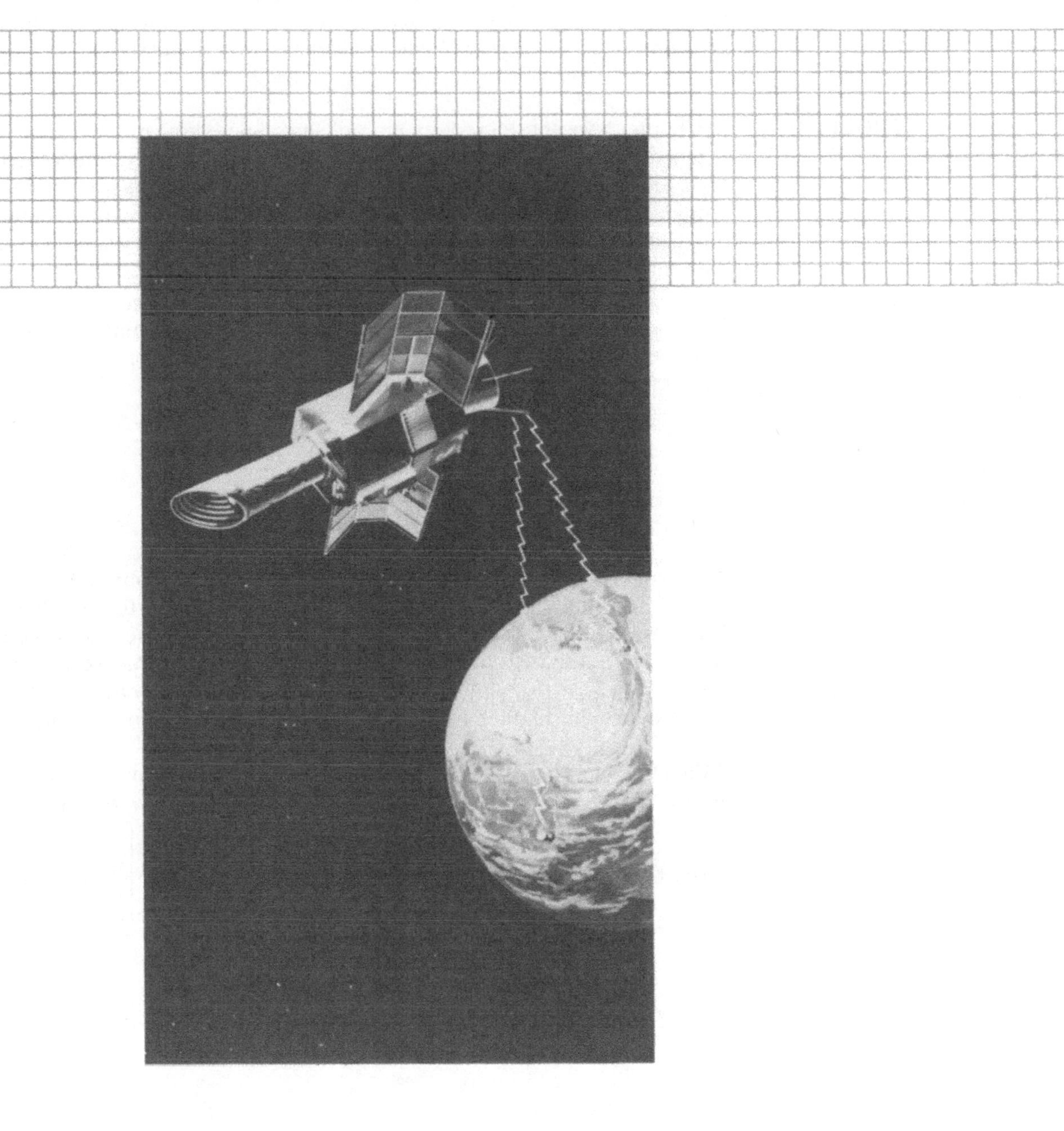

Artist's concept of the *International Ultraviolet Explorer* (IUE) showing how it can transmit and receive messages.

Matrices are an invaluable tool in space science, making it possible to organize, handle, and manipulate, with the aid of computers, large quantities of data. Most of the actual examples involving matrix algebra are too long and complex for inclusion here; however, by considering simplified examples, we can get some sense of the role of matrix algebra in this context.

PROBLEM 1. In Chapter 5 we considered some simple error-detecting binary codes for telemetry. A more complex system, the Hamming Code, will not only detect the presence of an error in a received message but will identify the erroneous bit in cases where a single error has occurred. If two bits are wrong, this fact will be detected but the locations of the errors will not be known. We use a very simple example to illustrate the method.

Suppose we have a "message" in the form of a four-bit binary string: that is, the message is in the form *abcd* where each of *a, b, c, d* is 0 or 1. The Hamming matrix for a message of this type is the 4×8 matrix H:

$$H = \begin{bmatrix} 0 & 0 & 0 & 0 & 1 & 1 & 1 & 1 \\ 0 & 0 & 1 & 1 & 0 & 0 & 1 & 1 \\ 0 & 1 & 0 & 1 & 0 & 1 & 0 & 1 \\ 1 & 1 & 1 & 1 & 1 & 1 & 1 & 1 \end{bmatrix}$$

The structure of the matrix is as follows: For a message containing 4 bits, we need $2^3 = 8$ columns and 4 rows. The binary numerals for 0 through 7, (written in 3-digit form as 000, 001, 010, . . .) are used, in order, as the first three entries in each column; the bottom entry is always 1. A Hamming matrix for a 5-bit message would need $2^4 = 16$ columns and 5 rows in order to represent the binary numerals for 0 through 15 (0000, 0001, . . . , 1111) followed by 1 in the columns.

If the message we wish to send is *abcd*, we need to use four additional parity bits, p_1, p_2, p_3, and p_4, and form a *message row vector* $M = [p_1\, p_2\, p_3\, a\, p_4\, b\, c\, d]$. The parity bits must be assigned so that the product $H \cdot M^T = \begin{bmatrix} 0 \\ 0 \\ 0 \\ 0 \end{bmatrix}$ in mod 2 arithmetic.

a. Find the conditions that p_1, p_2, p_3, and p_4 must satisfy so that $H \cdot M^T = \begin{bmatrix} 0 \\ 0 \\ 0 \\ 0 \end{bmatrix}$ in mod 2 arithmetic.

Solution:

$$H \cdot M^T = \begin{bmatrix} 0&0&0&0&1&1&1&1 \\ 0&0&1&1&0&0&1&1 \\ 0&1&0&1&0&1&0&1 \\ 1&1&1&1&1&1&1&1 \end{bmatrix} \cdot \begin{bmatrix} p_1 \\ p_2 \\ p_3 \\ a \\ p_4 \\ b \\ c \\ d \end{bmatrix} = \begin{bmatrix} p_4 + b + c + d \\ p_3 + a + c + d \\ p_2 + a + b + d \\ p_1 + p_2 + p_3 + a + p_4 + b + c + d \end{bmatrix} = \begin{bmatrix} 0 \\ 0 \\ 0 \\ 0 \end{bmatrix}.$$

So the conditions are $p_4 + b + c + d = 0$; $p_3 + a + c + d = 0$; $p_2 + a + b + d = 0$; $p_1 + p_2 + p_3 + a + p_4 + b + c + d = 0$.

b. Find the message row vector if the actual message is 0 1 1 0.

Solution: We have $a = 0, b = 1, c = 1, d = 0$. Substituting these values in the preceding equations in part (a) and solving in mod 2 gives $p_4 = 0, p_3 = 1, p_2 = 1, p_1 = 0$. The message row vector is then $M = [0\ 1\ 1\ 0\ 0\ 1\ 1\ 0]$.

c. The matrix $H \cdot M^{\mathrm{T}}$ is a column vector called the *syndrome vector S*. In the setting we are using, S will have four components. When a message is received, the syndrome vector is formed. If none of the bits of M was in error, the components of S will all be 0. If we find that $s_4 = 1$, we know that an error has occurred in transmission, and the binary number $s_1 s_2 s_3$ gives the number of the component of M which is wrong, where the components are numbered from the left, beginning with 0. If $s_4 = 0$ and one or more of s_1, s_2, s_3 is 1, then two bits of M are incorrect, but we do not know which two—the error is detectable but uncorrectable. If there are more than two errors, it is possible that they will be "corrected" incorrectly or not get detected.

Suppose the message [0 0 1 1 1 0 0 0] is received. Compute S and, if appropriate, correct the message.

Solution:

$$\begin{bmatrix} 0&0&0&0&1&1&1&1 \\ 0&0&1&1&0&0&1&1 \\ 0&1&0&1&0&1&0&1 \\ 1&1&1&1&1&1&1&1 \end{bmatrix} \cdot \begin{bmatrix} 0\\0\\1\\1\\1\\0\\0\\0 \end{bmatrix} = \begin{bmatrix} 1\\0\\1\\1 \end{bmatrix}$$

Since $s_4 = 1$, there is an error; $s_1 s_2 s_3 = 101_{\text{base } 2} = 5$, so the error is in position #5 (recall that the first position is #0) and the corrected message is [0 0 1 1 1 1 0 0].

In the last chapter we developed transformations from a spacecraft coordinate system to a reference system with the same origin when the spacecraft has performed a roll or a pitch or a yaw rotation. Matrix algebra is the natural tool to use to find the transformation in cases where the spacecraft performs a series of such rotations. This is developed in the next problem.

PROBLEM 2. Recall that in Problem 1 of Chapter 7, we showed that

$$x_{\mathrm{R}} = x \qquad x_{\mathrm{P}} = x \cos P - z \sin P \qquad x_{\mathrm{Y}} = x \cos Y + y \sin Y$$

$$y_{\mathrm{R}} = y \cos R + z \sin R \qquad y_{\mathrm{P}} = y \qquad y_{\mathrm{Y}} = y \cos Y - x \sin Y$$

$$z_{\mathrm{R}} = z \cos R - y \sin R \qquad z_{\mathrm{P}} = z \cos P + x \sin P \qquad z_{\mathrm{Y}} = z$$

where the uppercase R, P, Y are the angles of roll, pitch, and yaw respectively, the coordinates (x,y,z) are those of the reference system, and the subscripted coordinates are those of the spacecraft coordinate system after performance of the rotation designated by the subscript.

a. Express these transformations in matrix form $\begin{bmatrix} x_{sc} \\ y_{sc} \\ z_{sc} \end{bmatrix} = M \cdot \begin{bmatrix} x \\ y \\ z \end{bmatrix}$ where the subscript sc designates the spacecraft coordinate system, by finding M_R, M_P, M_Y, the matrices of roll, pitch, and yaw, respectively.

Solution: Expressing each set of transformations above in matrix form,

$$M_R = \begin{bmatrix} 1 & 0 & 0 \\ 0 & \cos R & \sin R \\ 0 & -\sin R & \cos R \end{bmatrix} \quad M_P = \begin{bmatrix} \cos P & 0 & -\sin P \\ 0 & 1 & 0 \\ \sin P & 0 & \cos P \end{bmatrix} \quad M_Y = \begin{bmatrix} \cos Y & \sin Y & 0 \\ -\sin Y & \cos Y & 0 \\ 0 & 0 & 1 \end{bmatrix}$$

b. If the spacecraft and reference systems are initially concurrent and the spacecraft performs in sequence a roll through angle R, a pitch through angle P, and a yaw through angle Y, then the transformation from reference system coordinates to spacecraft coordinates will be given by

$$\begin{bmatrix} x_{sc} \\ y_{sc} \\ z_{sc} \end{bmatrix} = M \cdot \begin{bmatrix} x \\ y \\ z \end{bmatrix} \quad \text{where } M = M_Y \cdot M_P \cdot M_R.$$

Find M if $R = 30°$, $P = 45°$, $Y = 60°$.

Solution:

$$M_R = \begin{bmatrix} 1 & 0 & 0 \\ 0 & \cos 30° & \sin 30° \\ 0 & -\sin 30° & \cos 30° \end{bmatrix} = \begin{bmatrix} 1 & 0 & 0 \\ 0 & \frac{\sqrt{3}}{2} & \frac{1}{2} \\ 0 & -\frac{1}{2} & \frac{3}{2} \end{bmatrix}$$

$$M_P = \begin{bmatrix} \cos 45° & 0 & -\sin 45° \\ 0 & 1 & 0 \\ \sin 45° & 0 & \cos 45° \end{bmatrix} = \begin{bmatrix} \frac{\sqrt{2}}{2} & 0 & -\frac{\sqrt{2}}{2} \\ 0 & 1 & 0 \\ \frac{\sqrt{2}}{2} & 0 & \frac{\sqrt{2}}{2} \end{bmatrix}$$

$$M_Y = \begin{bmatrix} \cos 60° & \sin 60° & 0 \\ -\sin 60° & \cos 60° & 0 \\ 0 & 0 & 1 \end{bmatrix} = \begin{bmatrix} \frac{1}{2} & \frac{\sqrt{3}}{2} & 0 \\ -\frac{\sqrt{3}}{2} & \frac{1}{2} & 0 \\ 0 & 0 & 1 \end{bmatrix}$$

$$M = M_Y \cdot M_P \cdot M_R = \begin{bmatrix} \frac{1}{2} & \frac{\sqrt{3}}{2} & 0 \\ -\frac{\sqrt{3}}{2} & \frac{1}{2} & 0 \\ 0 & 0 & 1 \end{bmatrix} \begin{bmatrix} \frac{\sqrt{2}}{2} & 0 & -\frac{\sqrt{2}}{2} \\ 0 & 1 & 0 \\ \frac{\sqrt{2}}{2} & 0 & \frac{\sqrt{2}}{2} \end{bmatrix} \begin{bmatrix} 1 & 0 & 0 \\ 0 & \frac{\sqrt{3}}{2} & \frac{1}{2} \\ 0 & -\frac{1}{2} & \frac{\sqrt{3}}{2} \end{bmatrix}$$

$$= \begin{bmatrix} \frac{1}{2} & \frac{\sqrt{3}}{2} & 0 \\ -\frac{\sqrt{3}}{2} & \frac{1}{2} & 0 \\ 0 & 0 & 1 \end{bmatrix} \begin{bmatrix} \frac{\sqrt{2}}{2} & \frac{\sqrt{2}}{4} & -\frac{\sqrt{6}}{4} \\ 0 & \frac{\sqrt{3}}{2} & \frac{1}{2} \\ \frac{\sqrt{2}}{2} & -\frac{\sqrt{2}}{4} & \frac{\sqrt{6}}{4} \end{bmatrix}$$

$$= \begin{bmatrix} \frac{\sqrt{2}}{4} & \frac{\sqrt{2}}{8} + \frac{3}{4} & -\frac{\sqrt{6}}{8} + \frac{\sqrt{3}}{4} \\ -\frac{\sqrt{6}}{4} & -\frac{\sqrt{6}}{8} + \frac{\sqrt{3}}{4} & \frac{3\sqrt{2}}{8} + \frac{1}{4} \\ \frac{\sqrt{2}}{2} & -\frac{\sqrt{2}}{4} & \frac{\sqrt{6}}{4} \end{bmatrix}$$

$$\doteq \begin{bmatrix} 0.35 & 0.93 & 0.13 \\ -0.61 & 0.13 & 0.78 \\ 0.71 & -0.35 & 0.61 \end{bmatrix}$$

c. The matrix M can be used to find the orientation of the spacecraft coordinate axes with respect to those of the reference system in terms of direction cosines. $\hat{X} = M \cdot \begin{bmatrix} 1 \\ 0 \\ 0 \end{bmatrix}$ is a column matrix whose elements are the direction cosines of the spacecraft x-axis with respect to the x-, y-, and z-axes of the reference system. Similarly, $\hat{Y} = M \cdot \begin{bmatrix} 0 \\ 1 \\ 0 \end{bmatrix}$ and $\hat{Z} = M \cdot \begin{bmatrix} 0 \\ 0 \\ 1 \end{bmatrix}$ produce column matrices whose elements are the direction cosines of the spacecraft y- and z-axes, respectively, with respect to the reference system. For the motion of part (*b*), find $\hat{X}$, $\hat{Y}$, and $\hat{Z}$, and, from these, the angles made by each of the spacecraft coordinate system axes with those of the reference system.

Solution:

$$\hat{X} = M \cdot \begin{bmatrix} 1 \\ 0 \\ 0 \end{bmatrix} = \begin{bmatrix} 0.35 \\ -0.61 \\ 0.71 \end{bmatrix} \doteq \begin{bmatrix} \cos 70^\circ \\ \cos (-52^\circ) \\ \cos 45^\circ \end{bmatrix}$$

The angles between the spacecraft x-axis and the x-, y-, and z-axes of the reference system are about 70°, −52°, and 45°, respectively.

$$\hat{Y} = M \cdot \begin{bmatrix} 0 \\ 1 \\ 0 \end{bmatrix} = \begin{bmatrix} 0.93 \\ 0.13 \\ -0.35 \end{bmatrix} \doteq \begin{bmatrix} \cos 22^\circ \\ \cos 83^\circ \\ \cos (-70^\circ) \end{bmatrix}$$

The angles between the spacecraft y-axis and the x-, y-, and z-axes of the reference system are about 22°, 83°, and −70°, respectively.

$$\hat{z} = M \cdot \begin{bmatrix} 0 \\ 0 \\ 1 \end{bmatrix} = \begin{bmatrix} 0.13 \\ 0.78 \\ 0.61 \end{bmatrix} \doteq \begin{bmatrix} \cos 83° \\ \cos 39° \\ \cos 52° \end{bmatrix}$$

The angles between the spacecraft z-axis and the x-, y-, and z-axes of the reference system are about 83°, 39°, and 52°, respectively.

Analyzing the light emitted from sources in space is a very important part of the astronomer's or space scientist's task. Some of these sources, such as the stars, are too far away for their shapes to be discernible; but others are close enough for the emitting volume to be made out—that is, light can be seen to come from separate parts of the volume—and such sources are said to be "spatially resolved." Among such sources are the solar atmosphere, glowing at temperatures ranging from 2500°C to well above a million degrees Celsius, depending on the particular location, and comet tails fluorescing under the Sun's radiation.

If such a source is transparent to its own radiation—that is, light emitted at any point within it can escape from the source volume without being scattered or reabsorbed—then an observer looking at a particular area of the surface of the source will see the sum of all the light emitted behind that area, in the "line of sight." The actual distribution of emitting intensity within the source, which in practice is always an unknown function of position, is not directly available to an outside observer.

However, when the source geometry is of an especially regular or simple shape, such as spherical or cylindrical, mathematical methods are available to "invert" the observed intensity data, thereby "reconstructing" the source.

In the next problem, we illustrate the basic idea with a very simple but concrete example in two dimensions.

PROBLEM 3. Consider a small checkerboard, three squares on a side, on which a few lighted candles have been placed in some squares at random, as shown in Fig. 8.1. If one looks down any row, the combined light of all the candles in that row will be seen; this combined light is simply the arithmetic sum of the separate candles in the row. Referring to the figure, if we look along row 1, the light of three candles will be seen; whereas, looking along row 3, we see the light of two candles. In similar fashion, one can look along a column or even along a diagonal.

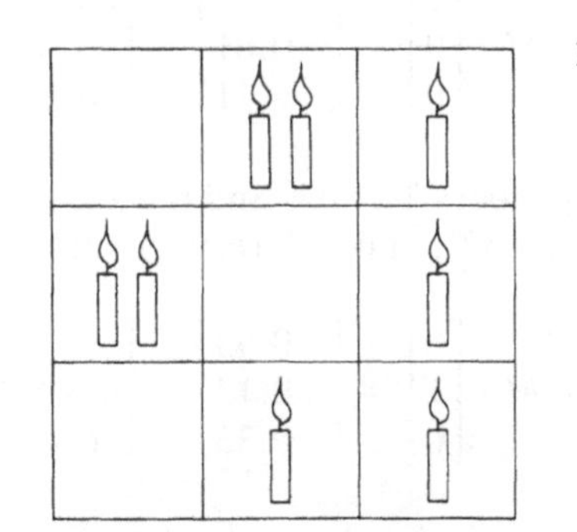

Fig. 8.1

To a "two-dimensional" observer in the plane of the checkerboard, this is, in fact, the only direct information available—the actual distribution of the candles on the checkerboard is unknown. This observer can, however, designate the number of candles in square (ij) as an unknown variable x_{ij} and proceed to set up a system of equations for these nine unknowns:

Row equations	*Column equations*
$x_{11} + x_{12} + x_{13} = 3$	$x_{11} + x_{21} + x_{31} = 2$
$x_{21} + x_{22} + x_{23} = 3$	$x_{12} + x_{22} + x_{32} = 3$
$x_{31} + x_{32} + x_{33} = 2$	$x_{13} + x_{23} + x_{33} = 3$

Since we need nine equations to solve for nine unknowns, we may look along three of the diagonals to get

$$x_{21} + x_{32} = 3$$

$$x_{11} + x_{22} + x_{33} = 1$$

$$x_{12} + x_{23} = 3$$

In this simple case, it is not difficult to solve the system by elimination; however, it is easy to see that a more general method is usually necessary.

a. Write a matrix equation for this system of linear equations.

Solution:

$$\begin{bmatrix} 1 & 1 & 1 & 0 & 0 & 0 & 0 & 0 & 0 \\ 0 & 0 & 0 & 1 & 1 & 1 & 0 & 0 & 0 \\ 0 & 0 & 0 & 0 & 0 & 0 & 1 & 1 & 1 \\ 1 & 0 & 0 & 1 & 0 & 0 & 1 & 0 & 0 \\ 0 & 1 & 0 & 0 & 1 & 0 & 0 & 1 & 0 \\ 0 & 0 & 1 & 0 & 0 & 1 & 0 & 0 & 1 \\ 0 & 0 & 0 & 1 & 0 & 0 & 0 & 1 & 0 \\ 1 & 0 & 0 & 0 & 1 & 0 & 0 & 0 & 1 \\ 0 & 1 & 0 & 0 & 0 & 1 & 0 & 0 & 0 \end{bmatrix} \cdot \begin{bmatrix} x_{11} \\ x_{12} \\ x_{13} \\ x_{21} \\ x_{22} \\ x_{23} \\ x_{31} \\ x_{32} \\ x_{33} \end{bmatrix} = \begin{bmatrix} 3 \\ 3 \\ 2 \\ 2 \\ 3 \\ 3 \\ 3 \\ 1 \\ 3 \end{bmatrix}$$

b. Use elementary row operations to find the source distribution for a case that produces the following matrix equation:

$$\begin{bmatrix} 1 & 1 & 1 & 0 & 0 & 0 & 0 & 0 & 0 \\ 0 & 0 & 0 & 1 & 1 & 1 & 0 & 0 & 0 \\ 0 & 0 & 0 & 0 & 0 & 0 & 1 & 1 & 1 \\ 1 & 0 & 0 & 1 & 0 & 0 & 1 & 0 & 0 \\ 0 & 1 & 0 & 0 & 1 & 0 & 0 & 1 & 0 \\ 0 & 0 & 1 & 0 & 0 & 1 & 0 & 0 & 1 \\ 0 & 0 & 0 & 1 & 0 & 0 & 0 & 1 & 0 \\ 1 & 0 & 0 & 0 & 1 & 0 & 0 & 0 & 1 \\ 0 & 1 & 0 & 0 & 0 & 1 & 0 & 0 & 0 \end{bmatrix} \cdot \begin{bmatrix} x_{11} \\ x_{12} \\ x_{13} \\ x_{21} \\ x_{22} \\ x_{23} \\ x_{31} \\ x_{32} \\ x_{33} \end{bmatrix} = \begin{bmatrix} 2 \\ 3 \\ 1 \\ 2 \\ 1 \\ 3 \\ 1 \\ 2 \\ 3 \end{bmatrix}$$

Solution: $x_{11} = 1;\quad x_{12} = 1;\quad x_{13} = 0;\quad x_{21} = 1;$

$x_{22} = 0;\quad x_{23} = 2;\quad x_{31} = 0;\quad x_{32} = 0;$

$x_{33} = 1$

Although it is certainly possible to solve part (b) manually, it is no doubt obvious that a computer solution is more desirable even in this vastly simplified context. Any of the commercially available programs to solve such matrix equations could be employed to produce the solution to part (b) or to discover that the solution to part (a) is not unique.

In practice, the physical radiating sources encountered are more complex in several ways: (a) they are continuous distributions rather than discrete ones, as in the example just treated; (b) they are three-dimensional sources; (c) they do not have simple geometric shapes; and (d) distant (astronomical) sources cannot usually be observed from a sufficient number of directions to obtain a complete set of emission data. What this means is that each observation must be modeled as an integral rather than a simple sum and the integrals are generally complicated expressions that are difficult to "invert." However, such inversions can be carried out for certain types of local radiating sources.

One recent example of this same technique in the medical field is Computer-Aided Tomography, or CAT scanning, in which X-ray radiation through a section of the human body is used to mathematically reconstruct a three-dimensional image of the section. For example, one kind of scanner measures the X-ray intensity that penetrates the portion of the body being imaged (such as the brain or the abdominal cavity). This scanner records the received radiation at 160 different positions in each scan direction; the entire unit is rotated one degree at a time around the head or abdomen, in a complete semicircle, to obtain 180 × 160, or 28 800, "sums." The computer then processes this information to produce a "picture" of a cross section of the organ by reconstructing the X-ray absorption in each square (or "pixel") of a 160 × 160 grid. The complexities enumerated in the foregoing paragraph also apply in this context, requiring the use of additional sophisticated mathematical techniques. However, the basic idea of the checkerboard model underlies this useful application.

CHAPTER NINE

CONIC SECTIONS

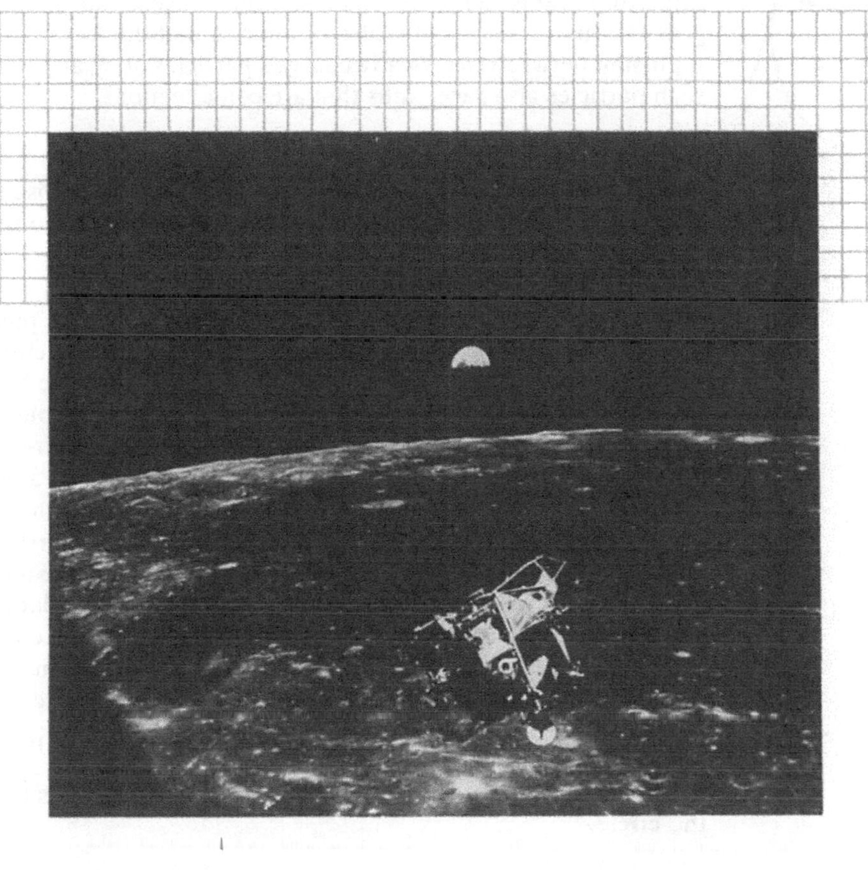

The *Apollo 11* lunar module photographed from the command and service module during rendezvous in lunar orbit with Earth rising above the lunar horizon.

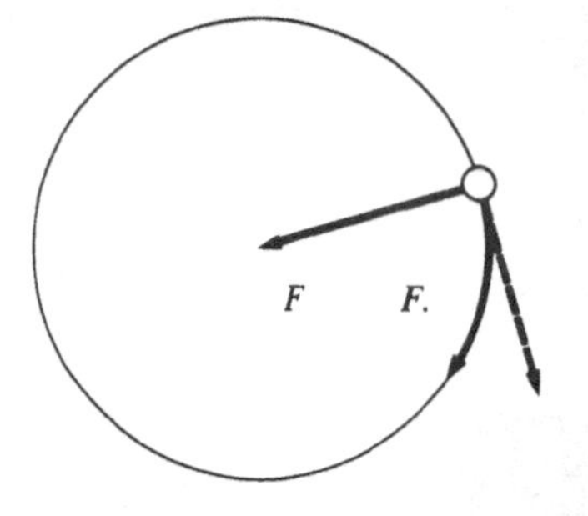

Fig. 9.1

The conic sections play a fundamental role in space science. As shown in the Appendix, any body under the influence of an inverse square law force (i.e., where force is inversely proportional to the square of distance) must have a trajectory that is one of the conic sections. In celestial mechanics the forces are gravitational; however, it is also of interest that the forces of attraction or repulsion between electrically charged particles obey an inverse square law, and such particles also have paths that are conic sections.

Telescopes with mirrors that are conic sections are also important in space technology because of their reflective properties. We shall close this chapter by considering the design of an X-ray telescope that requires two reflections in sequence from surfaces whose cross sections are conics.

In the analysis of orbits, where a celestial body, such as a planet, comet, meteor, star, or artificial satellite moves under gravitational attraction to a primary celestial body, the center of mass of the primary body is at one focus of the conic section along which the satellite moves. Because the simplest nontrivial conic section is the circle, we shall begin with a consideration of circular orbits. (The word "nontrivial" is included because a conic section could be a point or a pair of intersecting straight lines, if the sectioning plane passes through the cone's vertex.) Most of us understand from experience Newton's first law of motion, which states that an object in motion continues in a straight line unless it is acted on by some force. If we wish to make an object move in a circular path rather than in a straight line, we must give it a constant push toward the center. Thus a central, or centripetal, force is required. For example, when we tie a string to an object and whirl it in a circle, the pull of the string is the force that keeps the object in the circular path. If we represent the centripetal force by F_1, then $F_1 = \frac{mv^2}{r}$, where m is the mass of the object, v is its speed or velocity, and r is the radius of the circle.

When a spacecraft is moving in a circular orbit about any primary body, the force toward the center is supplied by the force of gravity F_2. According to Newton's law of universal gravitation, $F_2 = \frac{GMm}{r^2}$. In this equation, G is the constant of universal gravitation, assumed to be constant throughout the universe; M and m are the masses of any two bodies; and r is the distance between their centers of gravity. The physical situation, if the forces F_1 and F_2 are equal, is represented in Fig. 9.1.

The arrow toward the center represents the force of gravity, the dashed arrow represents the tangential velocity of the spacecraft, and the curved arrow indicates the circular path. (In rigorous use, velocity is a *vector* quantity, because it has both magnitude and direction, whereas speed, having magnitude only, is a *scalar* quantity. We will be using the symbol v for speed, the magnitude of the velocity vector.) Thus the force of gravity holds the body in the circular orbit.

If we set $F_1 = F_2$, we obtain $\frac{mv^2}{r} = \frac{GMm}{r^2}$. Solving for v gives us

$$v = \sqrt{\frac{GM}{r}}.$$

This simple equation enables us to find circular orbital velocities about any primary body, if M is the mass of the body and r is the radius of the orbit measured from the center of mass of the body. Because the value of GM is constant for any primary body, it is convenient to substitute its numerical value rather than to compute the value of the product for each individual problem. If the primary body is Earth, then $GM = 3.99 \times 10^{14}\ \text{m}^3/\text{s}^2$. Thus for bodies in circular orbits around Earth,

$$v_{\text{Earth}} = \sqrt{\frac{3.99 \times 10^{14}}{r}}\ \text{m/s}$$

where, of course, the distance r is expressed in meters.

PROBLEM 1. Most manned spacecraft in Earth orbit have been placed at altitudes of about 160 km or more because atmospheric drag at altitudes below this causes a rather rapid deterioration of the orbit. Find the velocity needed for a body to stay in Earth orbit at an altitude of 160 km.

Solution: Using the given equation,

$$v_{\text{Earth}} = \sqrt{\frac{3.99 \times 10^{14}}{(6380 + 160) \times 10^3}}\ \text{m/s}$$

$$= 10^5\sqrt{\frac{3.99}{654}}\ \text{m/s}$$

$$= 7.81 \times 10^3\ \text{m/s, or } 2.81 \times 10^4\ \text{km/h.}$$

PROBLEM 2. The formula for circular orbital velocity is quite general and can be applied to circular orbits about any primary body. G is a universal constant. We need only to change the value of M when we are concerned with another primary of different mass.

a. The mass of the Moon is approximately 0.012 times the mass M of Earth. Write a formula for finding circular orbital velocities about the Moon.

Solution: Multiplying the numerator in the previous equation by 0.012,

$$v_{\text{Moon}} = \sqrt{\frac{0.012 \times 3.99 \times 10^{14}}{r}}\ \text{m/s}$$

$$= \sqrt{\frac{4.8 \times 10^{12}}{r}}\ \text{m/s.}$$

b. During the *Apollo* flights the parking orbit for the command and service module about the Moon had an altitude of 110 km. The radius of the Moon is about 1740 km. Find the velocity in this orbit.

Solution:

$$v_{\text{Moon}} = \sqrt{\frac{4.8 \times 10^{12}}{(1740 + 110) \times 10^3}}\ \text{m/s}$$

$$= 10^3\sqrt{2.6}\ \text{m/s} = 1600\ \text{m/s},$$

or 5800 km/h.

PROBLEM 3. A synchronous Earth satellite is one that is placed in a west-to-east orbit over the equator at such an altitude that its period of revolution about Earth is 24 hours, the time for one rotation of Earth on its axis. Thus the orbital motion of the satellite is synchronized with Earth's rotation, and the satellite appears, from Earth, to remain stationary over a point on Earth's surface below. Such communication satellites as *Syncom*, *Early Bird*, *Intelsat*, and *ATS* are in synchronous orbits. Find the altitude and the velocity for a synchronous Earth satellite.

Solution: The velocity can be found from the equation for circular orbital velocity. It can also be found by dividing the distance around the orbit by the time required; that is, $v = \frac{2\pi r}{t}$. Because the two velocities are equal,

$$\frac{2\pi r}{t} = \sqrt{\frac{GM}{r}}$$

$$\left(\frac{2\pi r}{t}\right)^2 r = GM$$

$$r^3 = \frac{GMt^2}{4\pi^2}$$

$$r = \sqrt[3]{\frac{GMt^2}{4\pi^2}}.$$

It is apparent that $t = 24$ hours $= 86\,400$ seconds. Substituting the other values yields

$$r = \sqrt[3]{\frac{3.99 \times 10^{14} \times (86\,400)^2}{4 \times (3.14)^2}} = 10^7\sqrt[3]{75.4}$$

$$= 4.22 \times 10^7\ \text{m}, \qquad \text{or} \qquad 42\,200\ \text{km}$$

$$\text{Altitude} = 42\,200 - 6400\ \text{km} = 35\,800\ \text{km}$$

$$v = \frac{2 \times 3.14 \times 42\,200}{24} = 1.10 \times 10^4\ \text{km/hr}$$

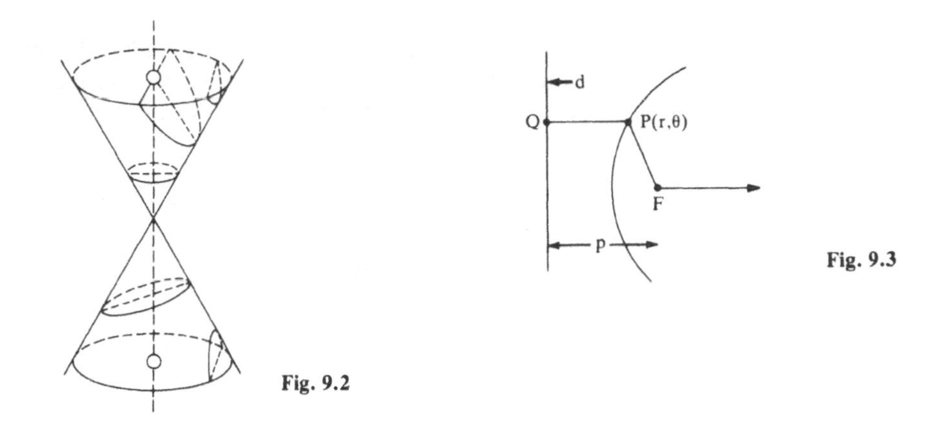

Fig. 9.2

Fig. 9.3

To understand orbits, we must know something of the nature and properties of the conic sections. They get their name, of course, from the fact that they can be formed by cutting or sectioning a complete right circular cone (of two nappes) with a plane. Any plane perpendicular to the axis of the cone cuts a section that is a circle. Incline the plane a bit, and the section formed is an ellipse. Tilt the plane still more until it is parallel to a ruling of the cone and the section is a parabola. Continue tilting until the plane is parallel to the axis and cuts both nappes, and the section is a hyperbola, a curve with two branches. It is apparent that closed orbits are circles or ellipses. Open or escape orbits are parabolas or hyperbolas (see Fig. 9.2).

Another way of classifying the conic sections is by means of their eccentricity. Let F be a fixed point (focus) and d a fixed line (directrix). For nonzero values of eccentricity e, a conic section may be defined as the locus of points such that the ratio of the distance PF to the distance from P to d is the constant e. The use of polar coordinates permits a unified treatment of the conic sections, and it is the polar coordinate equations of these curves that are used in celestial mechanics.

PROBLEM 4. Use the eccentricity definition above to show that the equation of a conic section in polar coordinates can be stated as $r = \dfrac{ep}{1 - e\cos\theta}$, where p is the distance between F and d, and the polar axis is perpendicular to, and pointing away from d, with the pole at F as shown in Fig. 9.3.

Solution: If Q is the foot of the perpendicular from P to d, and P has coordinates (r, θ), then, by definition,

$$e = \left|\frac{\text{PF}}{\text{PQ}}\right| = \frac{r}{p + r\cos\theta}$$

$$r = ep + er\cos\theta$$

$$r - er\cos\theta = ep$$

$$r(1 - e\cos\theta) = ep$$

$$r = \frac{ep}{1 - e\cos\theta}.$$

Fig. 9.4 shows a family of conics, each of which has directrix d and focus F_0, for different values of e. If a Cartesian coordinate system has origin F_0 and x-axis along the polar axis, the Cartesian equations of these conics have this form:

$$\text{ellipse:} \qquad \frac{(x-h)^2}{a^2} + \frac{y^2}{b^2} = 1$$

$$\text{parabola:} \qquad y^2 = q\,(x-h)$$

$$\text{hyperbola:} \qquad \frac{(x-h)^2}{a^2} + \frac{y^2}{b^2} = 1$$

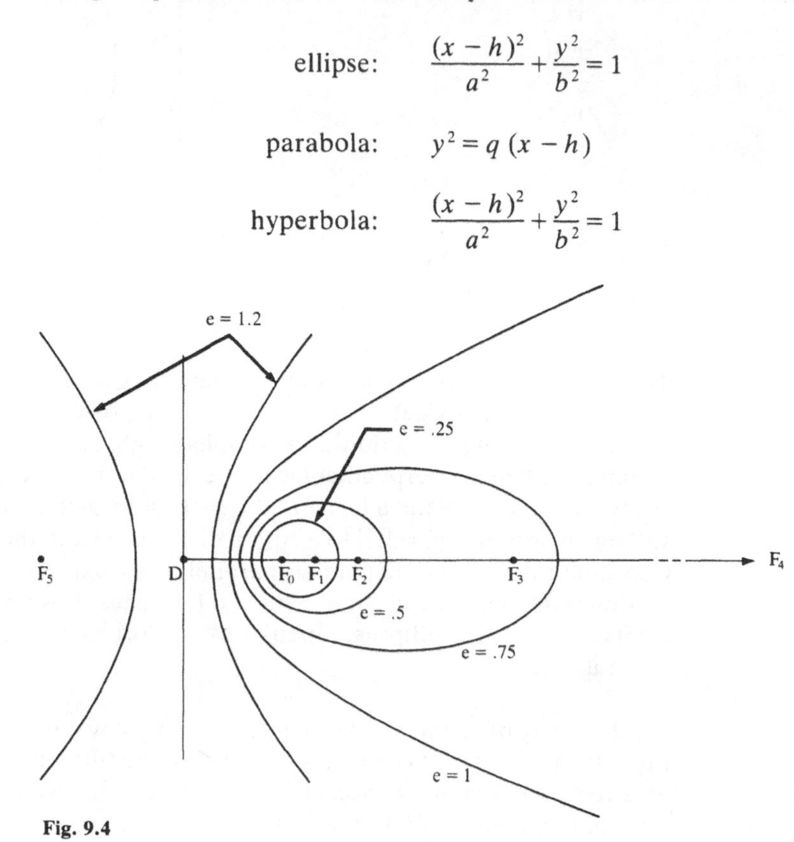

Fig. 9.4

PROBLEM 5. Show that the polar equation of Problem 4 can be transformed into the Cartesian equation of an ellipse if $0 < e < 1$; a parabola if $e = 1$; and a hyperbola if $e > 1$. Express the parameters h, a, b, or q, as appropriate, in terms of e and p.

Solution:

$$r = \frac{ep}{1 - e\cos\theta}$$

$$r = er\cos\theta + ep$$

If $e = 1$, then $r = er\cos\theta + ep$ becomes $r = r\cos\theta + p$. Since

$$r\cos\theta = x \quad \text{and} \quad r = \sqrt{x^2+y^2},\ \sqrt{x^2+y^2} = x + p.$$

Squaring,

$$x^2 + y^2 = x^2 + 2xp + p^2$$

$$y^2 = 2p\left(x + \frac{p}{2}\right),$$

so

$$h = -\frac{p}{2} \text{ and } q = 2p.$$

If $e \neq 1$, then $\qquad r = er\cos\theta + ep$ becomes

$$\sqrt{x^2 + y^2} = ex + ep$$

$$x^2 + y^2 = e^2x^2 + 2e^2px + e^2p^2$$

$$x^2(1 - e^2) - 2e^2px + y^2 = e^2p^2.$$

Dividing by $(1 - e^2)$ and completing the square, we get the following:

$$\left(x - \frac{pe^2}{1 - e^2}\right)^2 + \frac{y^2}{1 - e^2} = \frac{e^2p^2}{(1 - e^2)^2}$$

$$\frac{\left(x - \frac{pe^2}{1 - e^2}\right)^2}{\left(\frac{ep}{1 - e^2}\right)^2} + \frac{y^2}{\frac{e^2p^2}{1 - e^2}} = 1$$

If $0 < e < 1$, the denominator of the y^2 term is positive, and we have an ellipse with $h = \frac{pe^2}{1 - e^2}$, $a = \frac{ep}{1 - e^2}$, $b = \frac{ep}{\sqrt{1 - e^2}}$.

If $e > 1$, the denominator of the y^2 term is negative, so we may rewrite the equation as

$$\frac{\left(x - \frac{pe^2}{1 - e^2}\right)^2}{\left(\frac{ep}{e^2 - 1}\right)^2} - \frac{y^2}{\frac{e^2p^2}{e^2 - 1}} = 1,$$

and we have a hyperbola with

$$h = \frac{pe^2}{1 - e^2}, a = \frac{ep}{e^2 - 1}, b = \frac{ep}{\sqrt{e^2 - 1}}.$$

PROBLEM 6. Recall that for an ellipse $a^2 - b^2 = c^2$ and for a hyperbola $a^2 + b^2 = c^2$, where, in both cases, c is the distance between the center of the conic and a focus and a is the length of the semimajor axis. Show that the results of the preceding problem are consistent with this and that in both cases $e = c/a$.

Solution: For the ellipse,

$$c^2 = a^2 - b^2 = \frac{e^2p^2}{(1-e^2)^2} - \frac{e^2p^2}{1-e^2} = \frac{e^4p^2}{(1-e^2)^2}$$

$$c = \frac{e^2p}{1-e^2} = ea.$$

For the hyperbola,

$$c^2 = a^2 + b^2 = \frac{e^2p^2}{(e^2-1)^2} + \frac{e^2p^2}{e^2-1} = \frac{e^4p^2}{(e^2-1)^2}$$

$$c = \frac{e^2p}{e^2-1} = ea.$$

We see that for an ellipse, $2 \cdot c$ is the distance between the foci. Since $e = c/a$, if $c = 0$ we have $e = 0$; but if $c = 0$, the two foci coincide with the center and we have a circle rather than an ellipse. A circle can therefore be considered the conic section with eccentricity 0.

It is shown in the Appendix that the total energy E of a two-body gravitational system and the eccentricity e of the orbit of the less massive body (mass m) with respect to the more massive body (mass M) are related by

$$E = \frac{GMm(e^2-1)}{2ep}.$$

Since it is virtually impossible in the real world for the total energy to have a value that would result in $e = 0$ or $e = 1$ exactly, orbits that are exactly circles or exactly parabolas do not occur in nature. However, such orbits are of interest as limiting cases of actual trajectories. The energy equation of the Appendix,

$$\frac{1}{2}mv^2 - \frac{GMm}{r} = E = \frac{GMm(e^2-1)}{2ep},$$

provides the means to determine the velocity of an orbiting body at any point in its orbit.

PROBLEM 7. Solve the energy equation for v, and then express the velocity at any point in an orbit in terms of G, M, r and a, if needed (where a is defined as in problem 5), for each type of orbit.

Solution:

$$\frac{1}{2}mv^2 = \frac{GMm}{r} + \frac{GMm(e^2-1)}{2ep}$$

$$v^2 = GM\left(\frac{2}{r} + \frac{e^2-1}{ep}\right)$$

For an ellipse,

$$\frac{e^2-1}{ep} = -\frac{1}{a}, \text{ so}$$

$$v_e = \sqrt{GM\left(\frac{2}{r}-\frac{1}{a}\right)}.$$

For a circle, r = constant = a

$$v_c = \sqrt{GM\left(\frac{2}{r}-\frac{1}{r}\right)} = \sqrt{\frac{GM}{r}}.$$

(Recall that this was shown at the beginning of this chapter in the preliminary discussion of circular orbits.)

For a parabola, $e = 1$, and

$$v_p = \sqrt{GM\left(\frac{2}{r}\right)} = \sqrt{\frac{2GM}{r}}.$$

For a hyperbola,

$$\frac{e^2-1}{ep} = \frac{1}{a}, \text{ so that}$$

$$v_h = \sqrt{GM\left(\frac{2}{r}+\frac{1}{a}\right)}.$$

The minimum escape velocity of a rocket-borne space probe is the parabolic velocity $v_p = \sqrt{2GM/r}$. Velocities greater than this produce a hyperbolic orbit, and lesser velocities produce an elliptical orbit (or no orbit if too small).

Elliptical orbits are frequently analyzed in terms of orbit parameters, such as *apogee* and *perigee* distances. These distances are indicated in Fig. 9.5 by the letters A and P respectively. Before we discuss elliptical orbits, it will be necessary for us to avoid ambiguity by clarifying our terminology and mathematical notation. Most of us know from our reading of space events that in NASA news reports the point in an orbit nearest the surface of Earth is called the *perigee*, whereas the farthest point from the surface is called the *apogee*. These points are indicated by C and D, respectively, in Fig. 9.5. In common usage the word is used to refer to either the position of the point or the distance to the point.

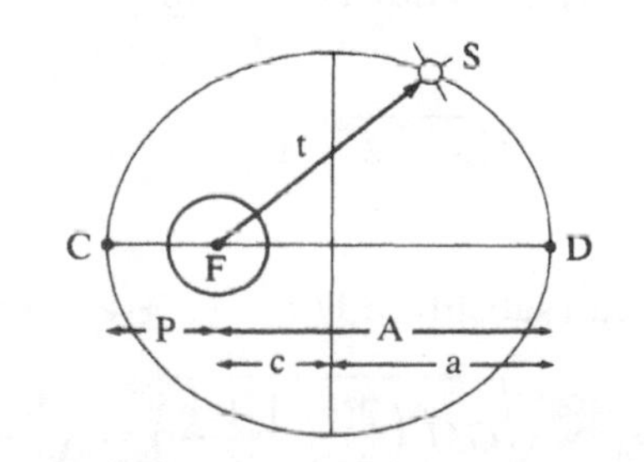

Fig. 9.5

However, usage is not uniform; some references state that the distances are measured, not from the surface of Earth, but from the center. In this article, we shall use distances measured from the center. The distances from the center to the perigee and the apogee will be indicated by P and A, respectively. In most discussions, the context will make this clear. If in any situation confusion could result, then distances from the surface, if used, will be called *perigee altitude* or *apogee altitude*, whereas distances from the center will be called *perigee radius* or *apogee radius*. Incidentally, the mathematics is simpler when distances are measured from the center.

PROBLEM 8. **a.** Express the distances A and P in terms of the semimajor axis a and the eccentricity e of an ellipse.

Solution: From Fig. 9.5,

$$A = a + c = a + ea = a(1 + e)$$

$$P = a - c = a - ea = a(1 - e).$$

b. Express the eccentricity of an elliptical orbit in terms of A and P.

Solution: The following relationships are apparent from Fig. 9.5:

$$a = \tfrac{1}{2}(A + P),$$

$$c = a - P = \tfrac{1}{2}(A + P) - P = \tfrac{1}{2}(A - P),$$

and

$$e = \frac{c}{a} = \frac{\frac{1}{2}(A - P)}{\frac{1}{2}(A + P)}$$

$$e = \frac{A - P}{A + P}$$

This formula is a quick and easy way of finding the eccentricity of an elliptical orbit. As a check, we note by inspection that $e = 0$ when $A = P$, which is the condition for a circular orbit.

PROBLEM 9. Derive formulas for v_A and v_P, the velocities at apogee and perigee, in terms of A or P, respectively, and e of the elliptical trajectory.

Solution: From Problem 7, the velocity of a body in an elliptical orbit at a distance r from the focus is

$$v = \sqrt{GM\left(\frac{2}{r} - \frac{1}{a}\right)}.$$

If $r = A = a(1 + e)$, we can substitute $1/a = (1 + e)/A$ and $r = A$ to get

$$v_A = \sqrt{GM\left(\frac{2}{A} - \frac{1 + e}{A}\right)} = \sqrt{\frac{GM}{A}(1 - e)}.$$

If $r = P = a(1 - e)$, by a similar substitution,

$$v_P = \sqrt{GM\left(\frac{2}{P} - \frac{1-e}{P}\right)} = \sqrt{\frac{GM}{P}(1+e)}.$$

These equations can be written in other ways as well, because numerous ways of expressing relationships among e, c, a, A, and P are possible. The particular form for the formulas reflects personal preference.

PROBLEM 10. Show that the velocities at apogee and perigee are inversely proportional to the distances from the center.

Solution: If we divide the equation for v_A by the equation for v_P (see Problem 9), we obtain

$$\frac{v_A}{v_P} = \sqrt{\frac{(1-e)P}{(1+e)A}} = \sqrt{\frac{a(1-e)P}{a(1+e)A}}$$

$$= \sqrt{\frac{P^2}{A^2}} = \frac{P}{A}.$$

Thus the velocity at perigee is inversely proportional to P, and so on. That is, when the orbital distance from the center of the primary body is small, the velocity at that point is large; when the distance is large, the orbital velocity is small. This result agrees with Kepler's second law of planetary motion, which states that a planet moves about the Sun in such a way that the radius vector from Sun to planet sweeps out equal areas in equal times.

PROBLEM 11. Derive a formula for the period of an elliptical orbit, given that the period of an elliptical orbit with semimajor axis a is the same as that for a circle with radius $r = a$.

Solution: Following the method used in Problem 3, we express the velocity in terms of the distance around the orbit and the time p required to make one transit of the orbit

$$v = \frac{2\pi r}{p}.$$

Also

$$v = \sqrt{\frac{GM}{r}}.$$

Then

$$\frac{2\pi r}{p} = \sqrt{\frac{GM}{r}}$$

$$\frac{(2\pi r)^2}{p^2} = \frac{GM}{r}$$

$$p^2 = \frac{(2\pi r)^2 r}{GM}$$

$$p = 2\pi\sqrt{\frac{r^3}{GM}}.$$

Because the period is the same when $r = a$, we may write

$$p = 2\pi\sqrt{\frac{a^3}{GM}}.$$

PROBLEM 12. An Earth satellite is placed in an elliptical orbit with perigee altitude of 160 km and apogee altitude of 16 000 km. Use 6380 km for the radius of Earth.

a. If injection is at perigee, what must be the injection velocity?

Solution: We first find the eccentricity as follows:

$$P = 6380 + 160 = 6540 \text{ km} \quad \text{or} \quad 6.54 \times 10^6 \text{ m}$$

$$A = 6380 + 16\,000 = 22\,380 \text{ km} \quad \text{or} \quad 2.24 \times 10^7 \text{ m}$$

By Problem 8,

$$c = \frac{22\,380 - 6540}{22\,380 + 6540} = \frac{15\,840}{28\,920} = 0.55.$$

By Problem 9,

$$v_P = \sqrt{\frac{3.99 \times 10^{14}}{6.54 \times 10^6}(1.55)} = 10^4\sqrt{0.9456}$$

$$= 9.72 \times 10^3 \text{ m/s}, \quad \text{or} \quad 3.50 \times 10^3 \text{ km/h}.$$

b. Find the speed at apogee.

Solution: By Problem 9,

$$v_A = \sqrt{\frac{3.99 \times 10^{14}}{2.24 \times 10^7}(1 - 0.55)} = \sqrt{1.78 \times 10^7 \times 0.45}$$

$$= 10^3\sqrt{8.02} \text{ m/s} = 2.83 \times 10^3 \text{ m/s}, \quad \text{or} \quad 1.02 \times 10^4 \text{ km/h}.$$

c. Find the period in this orbit.

Solution: From Problem 8,

$$a = \frac{22380 + 6540}{2} = 14460 \text{ km}, \quad \text{or} \quad 1.446 \times 10^7 \text{ m},$$

and, from Problem 11,

$$p = 2\pi\sqrt{\frac{(1.446 \times 10^7)^3}{3.99 \times 10^{14}}} \text{ s}$$

$$= (2\pi)(10^3)\sqrt{7.48} \text{ s}$$

$$= 17.2 \times 10^3 \text{ s}, \quad \text{or} \quad 4.77 \text{ h}.$$

PROBLEM 13. During the *Apollo* flights, the *Apollo* spacecraft and the third stage (SIVB) of the Saturn V launch vehicle were placed in a parking orbit 190 km above Earth. Find the velocity and period in this orbit.

Solution: Because $r = 6380 + 190 = 6570$ km, or 6.57×10^6 m, we find from Problem 7,

$$v_e = \sqrt{\frac{3.99 \times 10^{14}}{6.57 \times 10^6}} = 10^4\sqrt{0.6073}\ \text{m/s} = 7.79 \times 10^3\ \text{m/s}, \qquad \text{or} \qquad 2.8 \times 10^4\ \text{km/h}.$$

From Problem 11,

$$p = 2\pi\sqrt{\frac{(6.57 \times 10^6)^3}{3.99 \times 10^{14}}}\ \text{s} = 5300\ \text{s} = 1.47\ \text{h}.$$

PROBLEM 14. During the flight of *Apollo 11*, the SIVB stage was reignited and burned long enough to place the *Apollo* spacecraft on a trajectory to the Moon. At the end of the burn, the spacecraft had a velocity of about 3.90×10^4 km per hour at an altitude of 336 km. Was the *Apollo* spacecraft given escape velocity?

Solution: Using the results of Problem 7, the escape velocity equals

$$v_p = \sqrt{\frac{2GM}{r}} = \sqrt{\frac{2(3.99 \times 10^{14})}{(6380 + 336) \times 10^3}}\ \text{m/s}$$

$$= 10^4\sqrt{1.18}\ \text{m/s} = 1.09 \times 10^4\ \text{m/s}, \qquad \text{or} \qquad 3.92 \times 10^4\ \text{km/h}.$$

Thus the velocity imparted was about 200 km per hour less than escape velocity, thereby assuring a free return trajectory. That is, if the major propulsion systems failed, the spacecraft would be going slowly enough to be pulled around and oriented back toward Earth by lunar gravity, the attitude-control system being adequate to make needed course corrections.

PROBLEM 15. A spacecraft, as illustrated in Fig. 9.6, is in a circular orbit 800 km above Earth. The spacecraft must be transferred to a lower circular orbit 160 km above Earth. Compute the velocity changes needed at A and P to achieve this transfer.

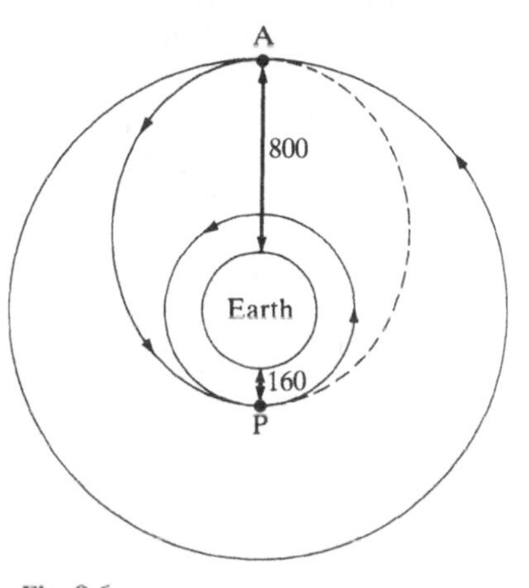

Fig. 9.6

Solution: We first find the eccentricity of the transfer orbit, which is, of course, an ellipse, with $A = 7180$ km and $P = 6540$ km.

$$e = \frac{7180 - 6540}{7180 + 6540} = \frac{640}{13720} = 0.047.$$

We then compare the velocities at A in the circular orbit and the elliptical orbit to find what changes must be made. Since GM has units m^3/s^2, we express A and P in meters. From Problem 7,

$$v_e = \sqrt{\frac{3.99 \times 10^{14}}{7.18 \times 10^6}}\ \text{m/s} = 7.45 \times 10^3\ \text{m/s}, \quad \text{or} \quad 2.68 \times 10^4\ \text{km/h},$$

and from Problem 9,

$$v_A = \sqrt{\frac{3.99 \times 10^{14}}{7.18 \times 10^6}(1 - 0.047)} = 7.28 \times 10^3\ \text{m/s}, \quad \text{or} \quad 2.62 \times 10^4\ \text{km/h}.$$

Therefore a propulsion engine on board the spacecraft must be fired long enough so that a retrothrust (opposite to the direction of motion) will slow down the spacecraft by 600 km per hour. The spacecraft will then leave the 800-km circular orbit and follow the elliptical transfer orbit, remaining in it indefinitely unless additional changes in velocity are made.

When the spacecraft reaches the point P, however, we want it to move from the elliptical orbit into the 160 km circular orbit. Therefore we must use the results of Problems 7 and 9 to investigate velocity changes at P.

$$v_e = \sqrt{\frac{3.99 \times 10^{14}}{6.54 \times 10^6}}\ \text{m/s} = 7.81 \times 10^3\ \text{m/s}, \quad \text{or} \quad 2.81 \times 10^4\ \text{km/h}$$

$$v_P = \sqrt{\frac{3.99 \times 10^{14}}{6.54 \times 10^6}(1.047)} = 7.99 \times 10^3\ \text{m/s}, \quad \text{or} \quad 2.88 \times 10^4\ \text{km/h}.$$

That is, a retrothrust must reduce velocity again, this time by about 700 km/h.

This method of transferring a spacecraft from one orbit to another is known as a Hohmann transfer, named after Walter Hohmann, city engineer of Essen, Germany, who published the method in 1925. There are many paths that could be used to move the spacecraft from the 800 km to the 160 km orbit. But the Hohmann-transfer ellipse, requiring only two short burns, is the most economical, taking the minimum amount of energy. Therefore this method is called a minimum-energy transfer. It has many applications.

PROBLEM 16. A satellite is placed into a synchronous orbit by a technique involving a Hohmann-transfer ellipse. We computed in Problem 3 that the altitude of such a satellite is about 35 800 km and its orbital speed is about 9370 km per hour. Fig. 9.7 suggests the details.

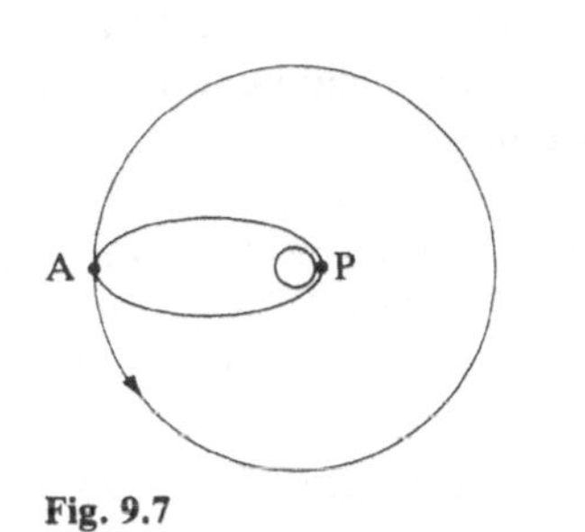

Fig. 9.7

We shall assume that injection is at the perigee point, which we shall place 160 km above Earth. Then obviously

$$P = 6380 + 160 = 6540 \text{ km, and}$$

$$A = 6380 + 35800 = 42180 \text{ km.}$$

We wish to find the velocity change needed at A.

Solution:

$$e = \frac{42180 - 6540}{42180 + 6540} = \frac{35640}{48720} = 0.732$$

$$v_P = \sqrt{\frac{3.99 \times 10^{14}}{6.54 \times 10^6}(1.732)} = 1.028 \times 10^4 \text{ m/s, or } 3.7 \times 10^4 \text{ km/h}$$

$$v_A = \sqrt{\frac{3.99 \times 10^{14}}{4.22 \times 10^7}(1 - 0.732)} = 1.59 \times 10^3 \text{ m/s, or } 5.73 \times 10^3 \text{ km/h.}$$

But the tangential velocity needed at point A is 9370 km per hour. Therefore the velocity of the satellite must be increased in the direction of Earth's rotation by 9370 − 5730 = 3640 km per hour. This extra push or kick would be provided by the firing of a motor on board the satellite, and the thrust and firing time must be such as to give the desired increment in velocity. Such a motor to be fired at apogee is called an *apogee motor*, and the thrust it provides is called an *apogee kick*.

The relative efficiency of using this method is easy to understand. Placing a heavy final stage of the launch vehicle at the synchronous altitude and then having a burn to give the entire assembly circular orbital velocity would take much fuel. Instead we send up to the synchronous altitude only a relatively light satellite and a small apogee motor. The numerical values used in this problem are merely illustrative. If the perigee altitude is higher or lower than the one we have assumed, all the other numbers are changed.

One more maneuver is needed to make the satellite synchronous. It now has a period equal to the time of Earth's rotation. However, the satellite will appear to be stationary over a given point only if it is in equatorial orbit. Unless corrections were made during launch, the plane of the orbit will be inclined to the plane of the equator. One method of solving this problem is to fire a motor at the precise instant when the satellite crosses the equator, adjusting the burn time and direction of thrust so that the vector sum of the burn velocity and the orbital velocity make the angle of inclination equal to zero.

PROBLEM 17. The first step in lunar orbit injection in the *Apollo 11* flight was to place the spacecraft in an elliptical orbit of 110 by 313 km, the low point—or *perilune* (corresponding to perigee for Earth)—being on the back side of the Moon.

a. Compute the velocity needed at perilune to inject the *Apollo* spacecraft into this orbit.

Solution: Using the data developed for lunar orbits in Problem 2,

$$P = 1740 + 110 = 1850 \text{ km}$$

$$A = 1740 + 314 = 2054 \text{ km}$$

$$e = \frac{2054 - 1850}{2054 + 1850} = \frac{204}{3904} = 0.052$$

$$v_p = \sqrt{\frac{4.8 \times 10^{12}}{1.85 \times 10^6}(1.052)} = 10^3\sqrt{2.73} = 1.65 \times 10^3 \text{ m/s},$$

$$\text{or } 5.95 \times 10^3 \text{ km/h}.$$

b. Find the period in this orbit.

Solution: Evidently $a = \frac{1}{2}(2054 + 1850)$ km, or 1.95×10^6 m

and

$$P = 2\pi\sqrt{\frac{(1.95 \times 10^6)^3}{4.8 \times 10^{12}}} \text{ s} = 2\pi(1.24) \times 10^2 \text{ s} = 782 \text{ s, or } 130 \text{ m}$$

PROBLEM 18. The lunar module descent orbit insertion during the *Apollo 11* mission began with a Hohmann transfer. The command and service (CSM) and lunar modules were in a circular orbit 110 km above the Moon. The lunar module was detached and its descent engine was fired to reduce velocity so that it would enter a 110-by-15-km lunar orbit. Find the reduction in velocity needed to achieve this orbit. The CSM remained in the 110 km parking orbit.

Solution: In this case, the change to the elliptical transfer orbit was made at apolune (corresponding to apogee for Earth).

$$A = 1740 + 110 = 1850 \text{ km or } 1.85 \times 10^6 \text{ m}$$

$$P = 1750 + 15 = 1755 \text{ km or } 1.755 \times 10^6 \text{ m}$$

$$e = \frac{1850 - 1755}{1850 + 1755} = \frac{95}{3605} = 0.026$$

$$v_A = \sqrt{\frac{4.8 \times 10^{12}}{1.85 \times 10^6}(1 - 0.026)} = 10^3\sqrt{2.526} = 1.59 \times 10^3 \text{ m/s},$$

or 5700 km/h.

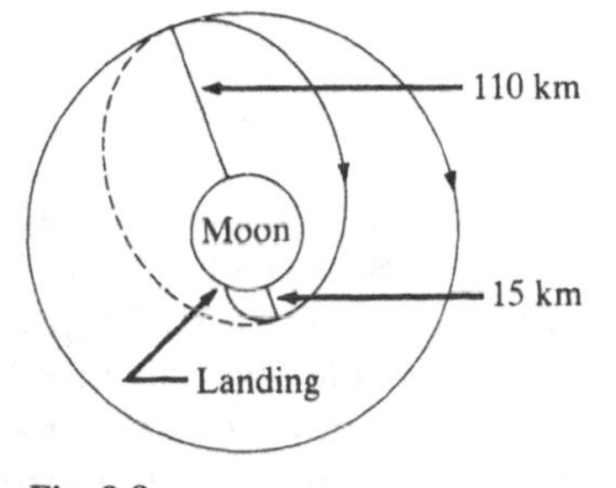

Fig. 9.8

We found in Problem 2 that the circular velocity in the 110-km orbit was 5800 km per hour. Thus the reduction in the velocity needed, achieved by a retroburn of the lunar module descent engine, was 100 km per hour. At perilune altitude of 15 km, several retroburns and attitude changes were made—both automatically, and manually by the pilot—causing the spacecraft to descend to the surface. If for any reason the descent from the 15-km perilune could not be made, the lunar module could have remained indefinitely in the elliptical transfer orbit until a rendezvous and docking with the CSM could be made. Thus this maneuver, which seemed so tricky and dangerous as we watched before our television sets, was actually a routine Hohmann transfer. The tricky maneuver, requiring some manual control, came when the powered descent to the lunar surface was made from the 15-km altitude.

We will conclude our discussion of orbits by considering the classic analysis known as Kepler's Problem, which in modern times makes use of high-speed computers to produce final results. It is the task of determining the exact position of a body in an elliptical orbit at any given time. Kepler, of course, was interested in establishing the nature of the planetary orbits around the Sun, but today the same analysis is used to predict the location of artificial satellites in their orbits around Earth.

We shall make use of a number of the relationships involving elliptical orbits already established. For the orbit illustrated in Figure 9.9,

$$r = \frac{ep}{1 - e\cos\theta} = \frac{a(1 - e^2)}{1 - e\cos\theta}$$

$$\text{FP} = p$$

$$\text{CP} = \text{CA} = a$$

$$\text{CF} = ae,$$

where F is one focus of the ellipse and the location of the primary body in the gravitational system; A, P, C are apogee, perigee, and center of the ellipse, respectively; and e is the eccentricity of the ellipse.

Kepler's Problem is stated in terms of the angle ν (called the *true anomaly* of the ellipse) between the Earth-perigee ray (FP) and the radius vector (FS), rather than the angle θ, as shown in Fig. 9.9. Since ν is the supplement of θ, the ellipse equation may be written in terms of ν as

$$r = \frac{a(1 - e^2)}{1 + e \cos \nu}.$$

Recall also that the rate at which the radius vector traces out the ellipse is not constant, but is in accordance with Kepler's first law: The radius vector sweeps out equal areas in equal time. This makes the task of expressing r and ν directly in terms of time extremely difficult. Kepler circumvented the problem by considering the projection of the ellipse on an "auxiliary circle" having the same center and passing through P and A as shown in Fig. 9.10. If a satellite is at S on the ellipse and Q is the foot of the perpendicular from S to AP, then S′ is the intersection of QS with the circle. Kepler defined three new quantities: the foreshortening factor $k = \mathrm{SQ}/\mathrm{S'Q}$; the eccentric anomaly $E = \angle \mathrm{S'CP}$; and the mean anomaly M, which is a fictitious angle through which an object would move at a uniform angular speed with respect to F. That is, $M = (\Delta t/T) \cdot 2\pi$ radians, where T is the time for one complete orbit and Δt is the time of interest. He then established the following relationships:

(1) $$k = \sqrt{1 - e^2}$$

(2) $$r = a(1 - e \cos E)$$

(3) $$\tan \frac{\nu}{2} = \sqrt{\frac{1 + e}{1 - e}} \tan \frac{E}{2}$$

(4) $$M = E - e \sin E$$

These will be derived in Problem 20.

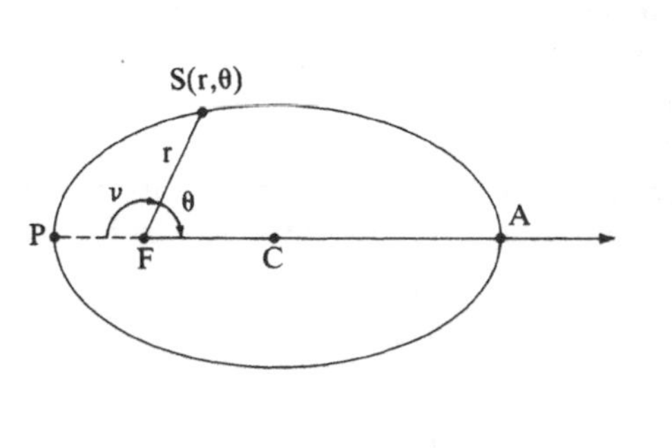

Fig. 9.9

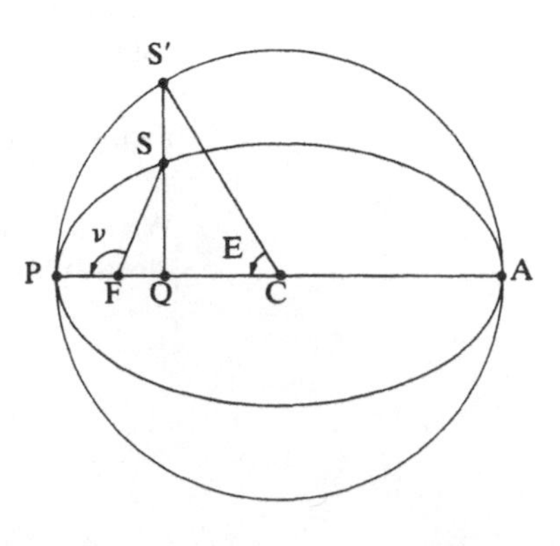

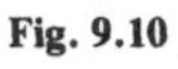
Fig. 9.10

In order to determine a satellite's position at any time, we must be able to compute r and ν Since in general, e and T (and therefore M) are known for an orbit, if equation (4) can be solved for E, then (2) and (3) will provide r and ν. But equation (4) is transcendental in E, so that no analytical solution is possible. This difficulty has been the core of many computational schemes generated by astronomers, mathematicians, and physicists.

High-speed computers now make a numerical, iterative solution both possible and feasible. The iterations would proceed as follows:

$$E_1 = 0$$

$$E_2 = M + e \sin E_1$$

$$E_3 = M + e \sin E_2$$

$$\vdots$$

$$E_{k+1} = M + e \sin E_k$$

The iteration continues as long as necessary to compute E to a desired accuracy (say 10^{-12} or 10^{-16}); in other words, when $|E_{k+1} - E_k| < 10^{-12}$, we can use $E = E_{k+1}$, if this is our desired level of accuracy. Since $E_{K+1} - E_k = e(\sin E_k - \sin E_{k-1})$, it can be shown that $|E_{k+1} - E_k| \leq e^{k-1}M$, so that the sequence E_k converges to the new value of E, since $e < 1$. This process is highly efficient for small values of e, and after a few iterations, it is usually found that the difference is within tolerance.

PROBLEM 19. **a.** Write a computer progream to perform the iteration outlined above, and then to use the value of E so found to compute r and ν, where E is found to an accuracy of 10^{-12}

```
]LIST

10  REM  ORBIT POSITION PROGRAM
20  REM  Q = ECCENTRICITY OF ORBI
     T
30  REM  A = SEMIMAJOR AXIS OF OR
     BIT
40  REM  P = PERIOD OF ORBIT
50  REM  T = TIME OF POSITION DET
     ERMINATION
100  PRINT "WHAT IS THE ECCENTRIC
     ITY OF THE ORBIT?": INPUT Q
110  PRINT "WHAT IS THE SEMIMAJOR
      AXIS LENGTH, ": PRINT " IN
     KILOMETERS?": INPUT A
120  PRINT "WHAT IS THE PERIOD OF
      THE ORBIT, ": PRINT " IN HO
     URS?": INPUT P
130  PRINT "HOW MANY HOURS AFTER
     PERIGEE IS THE "
140  PRINT "POSITION TO BE DETERM
     INED?": INPUT T
150 M = 2 * 3.1416 * T / P
160 E1 = 0
170  FOR J = 1 TO 20
180 E2 = M + Q *  SIN (E1)
190  IF  ABS (E2 - E1) < 10^(  -
     12) THEN 220
200 E1 = E2
210  NEXT J
215  PRINT "DESIRED ACCURACY NOT
     OBTAINED AFTER": PRINT "20 I
     TERATIONS"
220  PRINT J;" ITERATIONS WERE RE
     QUIRED TO": PRINT "ACHIEVE D
     ESIRED ACCURACY."
225  PRINT : PRINT
230 E = E2
240 R = A * (1 - Q *  COS (E))
245 R% = R:R = R%
250 W =  SQR ((1 + Q) / (1 - Q)) *
      TAN (E / 2)
260 NU = 2 *  ATN(W)
265 NU% = 100 * (NU + 0.005):NU =
     NU% / 100
270  PRINT " THE SATELLITE IS ";R
     ;" KM": PRINT " DISTANT FROM
      EARTH,"
280  PRINT " AND ITS ANOMALY IS
     ";NU;" RADIANS."
290  END
```

b. Use this program to find the position of the satellite discussed in Problem 12 one hour after it passes the perigee point in its orbit.

Solution: We had $e = 0.55$, $a = \frac{1}{2}(A + P) = \frac{1}{2}(22\ 380 + 6540)$ km $= 1.45 \times 10^4$ km, and $T = 4.77$ h. Running the program of part (a) with these values produces the following results:

```
]RUN
WHAT IS THE ECCENTRICITY OF THE ORBIT?
?0.55
WHAT IS THE SEMIMAJOR AXIS LENGTH,
 IN KILOMETERS?
?14500
WHAT IS THE PERIOD OF THE ORBIT,
 IN HOURS?
?4.77
HOW MANY HOURS AFTER PERIGEE IS THE
POSITION TO BE DETERMINED?
?1
12 ITERATIONS WERE REQUIRED TO
ACHIEVE DESIRED ACCURACY.

  THE SATELLITE IS 16670 KM
  DISTANT FROM EARTH,
  AND ITS ANOMALY IS 2.37 RADIANS.
```

PROBLEM 20. The four relationships of Kepler's Problem can be established using the geometry and trigonometry of Fig. 9.10.

a. Show that $k = \sqrt{1 - e^2}$

Solution: k was defined as $k = \mathrm{SQ}/\mathrm{S'Q}$. From Fig. 9.10, we have

$$\mathrm{SQ} = r \sin \nu,$$

and

$$\mathrm{S'Q} = \sqrt{(\mathrm{S'C})^2 - (\mathrm{CQ})^2} = \sqrt{a^2 - (\mathrm{CF} - \mathrm{QF})^2}$$
$$= \sqrt{a^2 - (ae + r\cos\nu)^2}.$$

Then

$$k = \frac{r \sin \nu}{\sqrt{a^2 - (ae + r\cos\nu)^2}} = \sqrt{\frac{r^2(1 - \cos^2\nu)}{a^2 - (ae + r\cos\nu)^2}}.$$

Substituting $r = \dfrac{a(1 - e^2)}{1 + e\cos\nu}$ and simplifying produces, after some labor, the result $k = \sqrt{1 - e^2}$.

b. If a rectangular coordinate system is placed on Fig. 9.10 with origin at F and positive x-axis along the polar axis, express the rectangular coordinates of S in terms of E, a, and e. Then use the fact that S can also be given by $(-r\cos\nu, r\sin\nu)$ to show that $r = a(1 - e\cos E)$.

Solution: Since CF = ae and S′C = a, the x-coordinate of S is QF = CF − CQ = $ae - a\cos E$. The y-coordinate is

$$\begin{aligned} SQ &= k\,S'Q \\ &= \sqrt{1-e^2}\,(a \sin E). \end{aligned}$$

So the coordinates of S are $(ae - a\cos E, \sqrt{1-e^2}\,a\sin E)$.

$$\begin{aligned} \text{Now} \quad r^2 &= r^2\cos^2\nu + r^2\sin^2\nu = (-r\cos\nu)^2 + (r\sin\nu)^2 \\ &= (ae - a\cos E)^2 + (\sqrt{1-e^2}\,a\sin E)^2 \\ &= a^2e^2 - 2a^2e\cos E + \underbrace{a^2\cos^2 E + a^2\sin^2 E}_{\downarrow} - a^2e^2\sin^2 E \\ &= a^2e^2 - 2a^2e\cos E + a^2 - a^2e^2(1-\cos^2 E) \\ &= a^2 - 2a^2e\cos E + a^2e^2\cos^2 E \\ &= a^2(1 - 2e\cos E + e^2\cos^2 E) \\ &= a^2(1-e\cos E)^2. \end{aligned}$$

$$\text{So} \quad r = a(1 - e\cos E).$$

c. Use (b) and the identity

$$\tan\frac{\theta}{2} = \sqrt{\frac{1-\cos\theta}{1+\cos\theta}}$$

to show that

$$\tan\frac{\nu}{2} = \sqrt{\frac{1+e}{1-e}}\tan\frac{E}{2}.$$

Solution: Since $r = a(1 - e\cos E)$ and $-r\cos\nu = a(e - \cos E)$,

$$\cos\nu = \frac{\cos E - e}{1 - e\cos E}.$$

Then

$$1 - \cos\nu = \frac{1 - e\cos E - \cos E + e}{1 - e\cos E} = \frac{(1+e)(1-\cos E)}{1 - e\cos E},$$

and

$$1 + \cos\nu = \frac{1 - e\cos E + \cos E - e}{1 - e\cos E} = \frac{(1-e)(1+\cos E)}{1 - e\cos E}.$$

Then

$$\tan\frac{\nu}{2} = \sqrt{\frac{1-\cos\nu}{1+\cos\nu}} = \sqrt{\frac{1+e}{1-e}}\sqrt{\frac{1-\cos E}{1+\cos E}} = \sqrt{\frac{1+e}{1-e}}\tan\frac{E}{2}.$$

d. As the satellite moves in its orbit so that the radius vector sweeps out equal areas of the ellipse in equal times, FS′ sweeps out equal areas of the auxiliary circle in equal times. If the area enclosed by FP, FS′, and the arc S′P is swept out in time Δt, and T is the time for the satellite to traverse the entire ellipse, then this area is given by $\Delta t/T\,(\pi a^2)$. Recall that M is $(\Delta t/T)(2\pi)$, and use the geometry of Fig. 9.10 to show that $M = E - e\sin E$.

Solution:

$$\text{Area FPS}' = \text{area of sector CSP}' - \text{area of CFS}'$$

$$\frac{\Delta t}{T}\pi a^2 = \frac{1}{2}a^2 E - \frac{1}{2}(ae)(a\sin E)$$

$$2\pi\frac{\Delta t}{T} = E - e\sin E$$

$$M = E - e\sin E$$

Among the first telescopes used to explore the heavens were those based on the reflective properties of paraboloidal mirrors. It is the fact that all light striking such a mirror in the direction parallel to the axis of the paraboloid is reflected to the focus that provides the light-gathering capacity of the telescope. The reflective properties of ellipsoidal and hyperboloidal surfaces are also important. In both cases, light striking the surface in a direction toward or away from one focus is reflected in a direction either away from or toward the other focus. These properties are illustrated in Fig. 9.11.

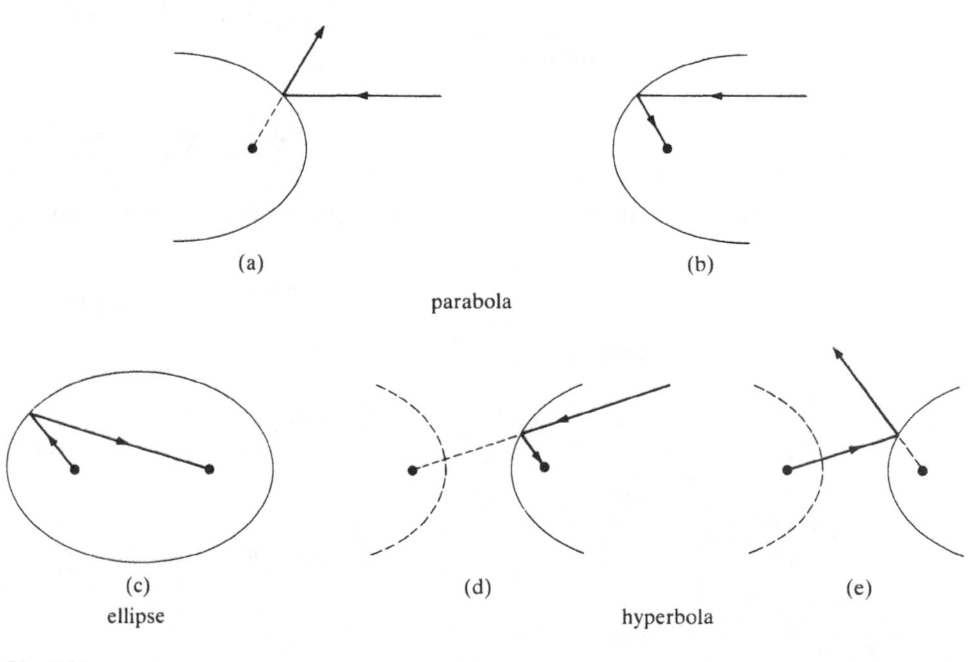

Fig. 9.11

PROBLEM 21. The technique of ray tracing is used in the design of optical instruments. One class of such instruments owes its focusing properties to the law of reflection. This law states that the angle between the incident ray and the reflecting surface must equal the angle between the emergent ray and the reflecting surface. In ray tracing, equations are written for the lines containing incident and emergent rays of electromagnetic radiation. Use this technique to prove the reflective property of the parabola, given that the slope of the tangent at the point (x_0, y_0) on the parabola $y^2 = 4px$ is $2p/y_0$. (The slope of the tangent at any point on a conic section graph will be derived in Chapter 10.)

Solution: We must show that an incident ray parallel to the axis is reflected through the focus. The geometry of the reflection is shown in Fig. 9.12; since the lines TI and PF are parallel, and since the angle of incidence equals the angle of reflection, we have that triangle FTP is isosceles and so $\phi = 2\theta$. The equation of the line containing the incident ray is $y = y_0$. The equation of the line containing the reflected ray is $(y - y_0) = \tan\phi\,(x - x_0) = \tan 2\theta\,(x - x_0)$.

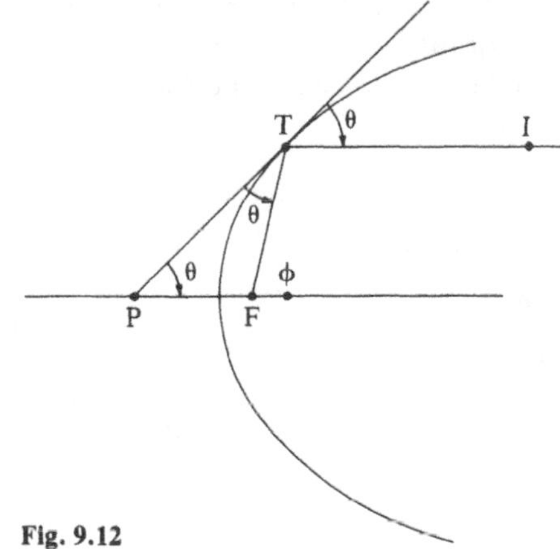

Fig. 9.12

Since

$$\tan 2\theta = \frac{2\tan\theta}{1 - \tan^2\theta} \text{ and } \tan\theta = \frac{2p}{y_0},$$

$$\tan 2\theta = \frac{4p}{y_0} \Big/ \left(1 - \frac{4p^2}{y_0^2}\right) = \frac{4py_0}{y_0^2 - 4p^2}.$$

But

$$y_0^2 = 4px_0$$

so

$$\tan 2\theta = \frac{4py_0}{4px_0 - 4p^2}$$

$$= \frac{y_0}{x_0 - p}.$$

Substituting the slope into the equation of the line,

$$y - y_0 = \frac{y_0}{x_0 - p}(x - x_0)$$

$$y = \frac{y_0(x - x_0)}{x_0 - p} + y_0 .$$

If $x = p$, $y = 0$, so this line passes through the focus.

Special instruments have been designed to study the electromagnetic radiation of stars and other astronomical sources in wavelengths outside the visible region. If X-rays are to be reflected, the incoming rays must form a very small angle (grazing angle) with respect to the reflecting surface; otherwise the X-rays are simply absorbed. However, with grazing angle incidence, incoming rays that are not parallel to the axis are not focused at all (making it impossible to form an image of a source that is not a point) unless an even number of reflections is used. The X-ray telescope on the *High Energy Astronomy Observatory* (HEAO) satellite was therefore designed to use two reflections from conic section surfaces. Fig. 9.13 shows some of the possibilities that were considered. Notice that in each, the focus of the paraboloid coincides with one focus of the other conic.

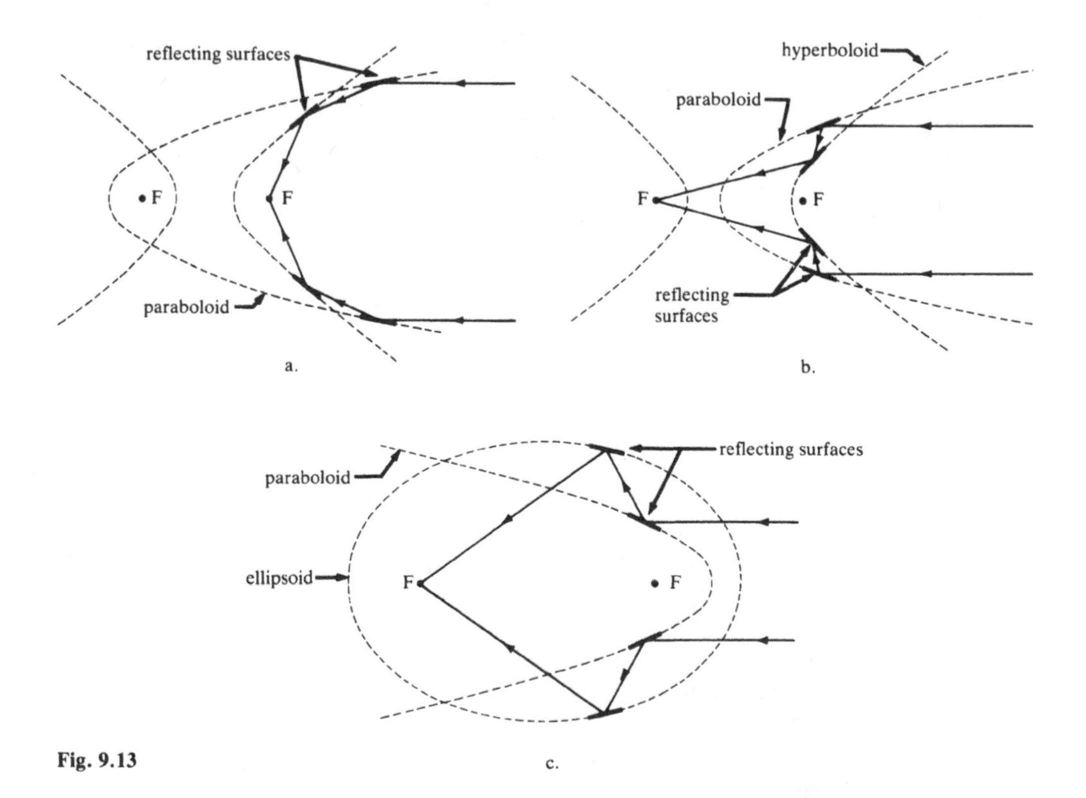

Fig. 9.13

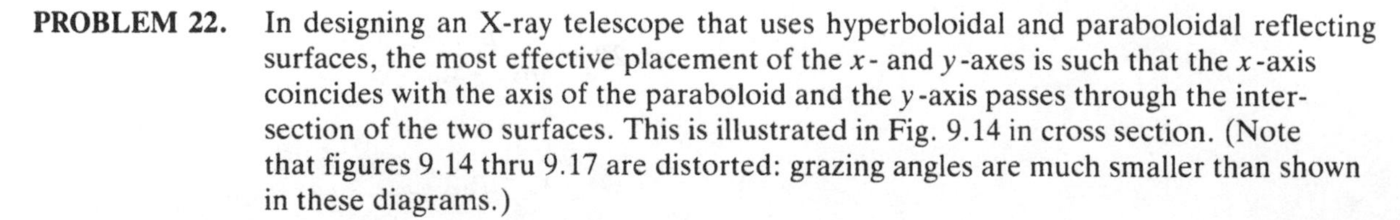

PROBLEM 22. In designing an X-ray telescope that uses hyperboloidal and paraboloidal reflecting surfaces, the most effective placement of the x- and y-axes is such that the x-axis coincides with the axis of the paraboloid and the y-axis passes through the intersection of the two surfaces. This is illustrated in Fig. 9.14 in cross section. (Note that figures 9.14 thru 9.17 are distorted: grazing angles are much smaller than shown in these diagrams.)

Fig. 9.14

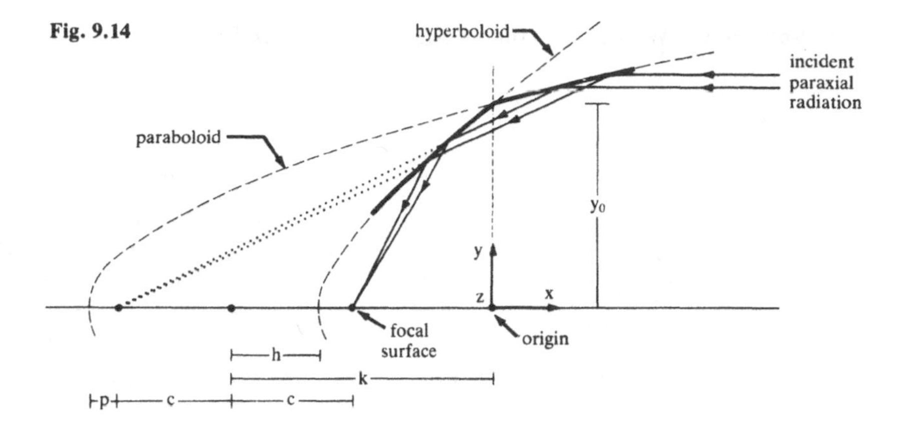

a. If p is the distance between the vertex and focus of the parabola in Fig. 9.14; c, the distance between the center of the hyperbola and each of its foci; h, the distance between the center of the hyperbola and each of its vertices; and k, the distance between the center of the hyperbola and the origin, find the equations of the two conic sections in this coordinate system.

Solution: For the parabola, the vertex is at $(-(p + c + k), 0)$ and the focus-vertex distance is p, so the equation is

$$y^2 = 4p\,(x + p + c + k),$$

For the hyperbola, the center is at $(-k, 0)$; the role of the parameter a in the standard equation is taken by h and that of the parameter b by $\sqrt{c^2 - h^2}$, so the equation is

$$\frac{(x + k)^2}{h^2} - \frac{y^2}{c^2 - h^2} = 1.$$

b. Fig. 9.15 shows the ray paths, which form angles α and β with respect to the x-axis, and the tangents to the parabola and hyperbola, which form angles θ and ϕ respectively, with respect to the x-axis. Experience in this field has shown that when successive reflections take place, surface reflection efficiency is maximum when an incoming ray parallel to the axis strikes each reflecting surface at about the same angle. Show that this condition, together with the fact that the angle of incidence equals the angle of reflection, means that $\beta = 2\theta$, $\phi = 3\theta$, $\alpha = 4\theta$.

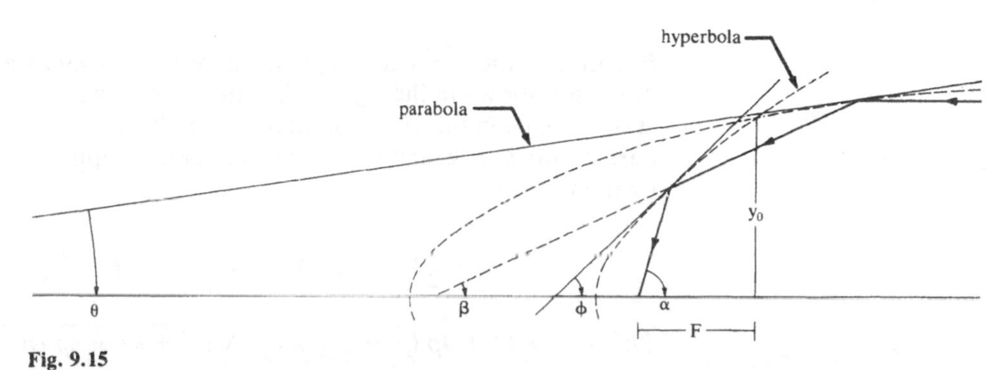

Fig. 9.15

Solution: We number the angles 1, 2, 3, 4, 5 in Fig. 9.15, as shown in Fig. 9.16.

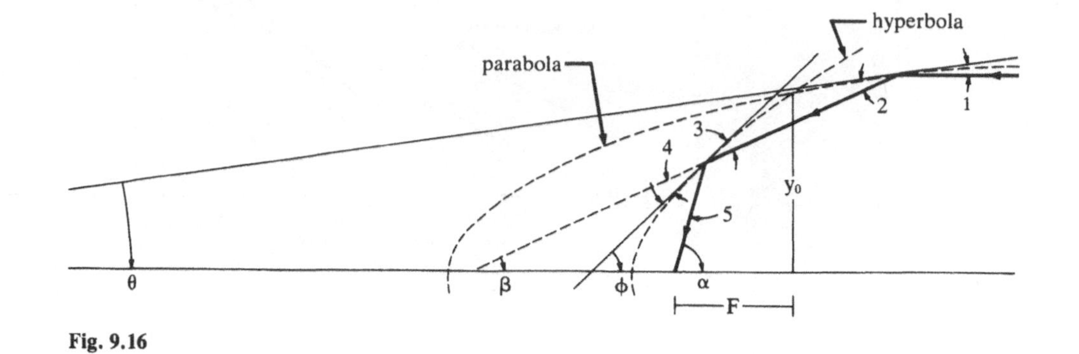

Fig. 9.16

We see that

$$\angle 1 = \angle 2 \text{ and } \angle 3 = \angle 5 \text{ } (\angle \text{ of incidence} = \angle \text{ of reflection});$$

$$\angle 3 = \angle 4 \text{ (vertical angles);}$$

$$\angle 1 = \theta \text{ (incoming ray is parallel to axis);}$$

$$\angle 2 = \angle 3 \text{ (for maximum reflection efficiency).}$$

So

$$\angle 1 = \angle 2 = \angle 3 = \angle 4 = \angle 5 = \theta.$$

Now, since an exterior angle of a triangle is equal to the sum of the nonadjacent interior angles,

$$\beta = 2\theta;$$

$$\phi = \beta + \theta = 3\theta;$$

$$\alpha = \phi + \theta = 4\theta.$$

c. In designing the surface of the X-ray telescope, the designer must be able to express the parameters p, c, k, and h of part (a) in terms of two initial design parameters F and y_0 of the instrument, where F is the distance along the x-axis between the origin and the focus of the hyperbola, and y_0 is the distance along the y-axis between the origin and the point of intersection to the parabola and hyperbola. (These are shown in Fig. 9.15.)

Recall that under grazing-angle incidence, θ is a very small angle (this is definitely not shown in the figure; the angle occurs where these lines finally intersect). It is also true in this situation that p is small compared to F. Show that the parameters p, c, k, and h can be given, at least approximately, in terms of F and y_0 by the following

$$c = \frac{1}{2}F \qquad k = \frac{3}{2}F \qquad p = \frac{y_0^2}{8F}$$

$$2h^2 = c^2 + k^2 + 4p(p + c + k) - \sqrt{(c^2 + k^2 + 4p(p + c + k))^2 - 4c^2k^2}$$

Solution: Let (x_1, y_1) be the point on the hyperbola where the second reflection takes place, and let $(0, y_2)$ be the y-intercept of the line containing the ray after the second reflection. (See Figure 9.17.) Then $\tan \alpha = y_2/F$ and $\tan \beta = y_1/(c + k) = y_1/(2c + F)$, since $F = k - c$. Because θ is such a small angle, α and β are also small. This suggests the following approximations:

$$y_2 \doteq y_0 \doteq y_1; \tan \alpha \doteq \alpha; \tan \beta \doteq \beta.$$

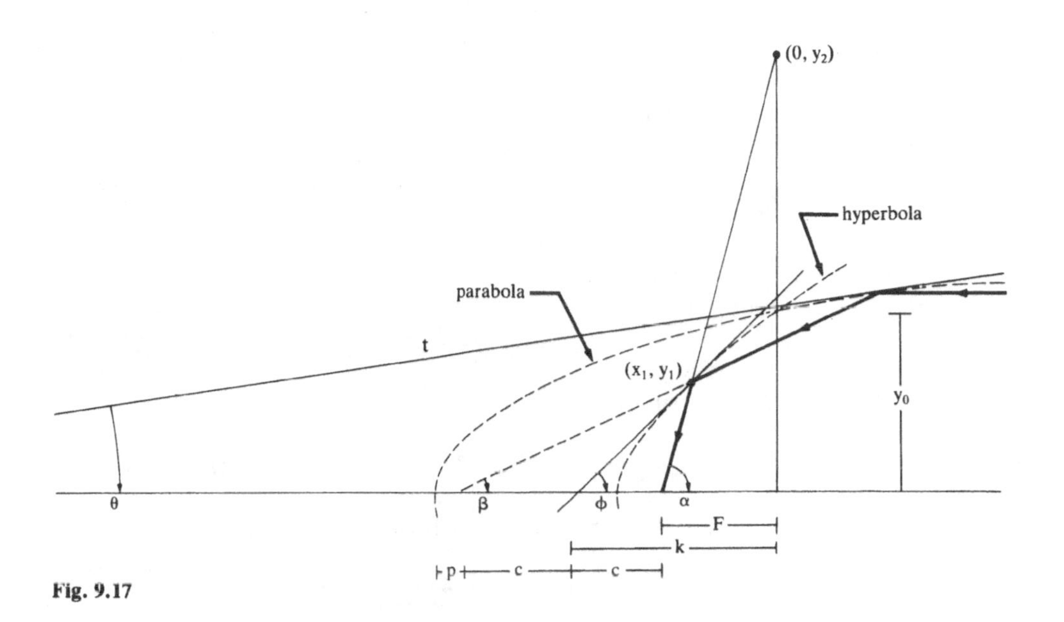

Fig. 9.17

(Fig. 9.17 does not show this because the angles are not small enough. The reader is encouraged to imagine how the figure would change if angles θ, β, ϕ, and α shrink.)

Since $\alpha = 2\beta$, we get

$$\frac{y_2}{F} \doteq \frac{2y_1}{2c + F},$$

and if $y_2 \doteq y_1$, then

$$2F \doteq 2c + F, \qquad \text{or} \qquad F \doteq 2c.$$

So

$$c \doteq \frac{1}{2}F, \qquad \text{and} \qquad k = F + c \doteq \frac{3}{2}F.$$

The parabola has equation $y^2 = 4p\,(x + p + c + k)$. Since $y = y_0$ when $x = 0$, and since $c + k \doteq 2F$,

$$y_0^2 \doteq 4p\,(p + 2F) = 4p^2 + 8Fp.$$

If p is small compared to F, the term in p^2 may be neglected giving $p \doteq y_0^2/(8F)$; h can be expressed exactly in terms of p, c, and k by observing that $(0, y_o)$ is on both the parabola and the hyperbola. This means that

$$y_0^2 = 4p\,(p + c + k)$$

and

$$\frac{k^2}{h^2} - \frac{y_0^2}{c^2 - h^2} = 1.$$

Then

$$y_0^2 = (c^2 - h^2)\left(\frac{k^2}{h^2} - 1\right) = 4p\,(p + c + k);$$

so

$$c^2k^2 - h^2(c^2 + k^2) + h^4 = 4p\,(p + c + k)h^2,$$

or

$$h^4 - h^2(c^2 + k^2 + 4p\,(p + c + k)) + c^2k^2 = 0.$$

Using the quadratic formula,

$$h^2 = \frac{c^2 + k^2 + 4p\,(p + c + k) \pm \sqrt{(c^2 + k^2 + 4p\,(p + c + k))^2 - 4c^2k^2}}{2}.$$

The positive sign before the radical is discarded, since it produces a physically unrealistic value of h larger than c, so

$$2h^2 = c^2 + k^2 + 4p\,(p + c + k) - \sqrt{(c^2 + k^2 + 4p\,(p + c + k))^2 - 4c^2k^2}.$$

This analysis produces an initial set of parameters p, c, k, and h. A ray-tracing computer program, based on the principles discussed in Problem 21, is then used to check the actual focusing capabilities of a hypothetical instrument with these specifications. The use of the computer with such a program makes it possible to refine the values of the parameters for best focus under desired conditions. Actual preparation of the reflecting surface is also controlled by computer-driven machinery once the optimal values of the parameters are established.

CHAPTER TEN

CALCULUS

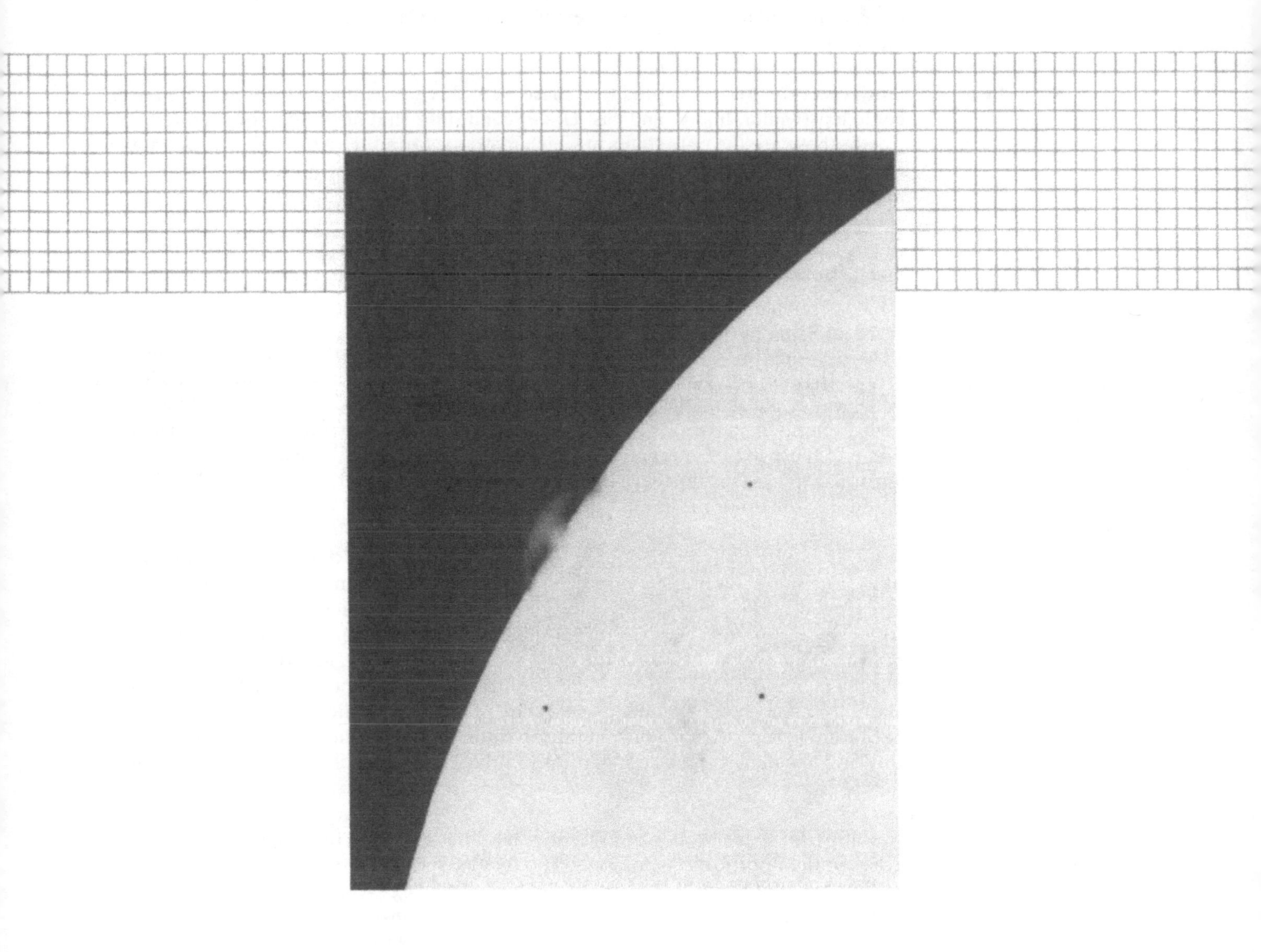

Photograph of an active volcanic eruption on Jupiter's satellite Io taken on March 4, 1979, by *Voyager 1*.

Although calculus is used extensively in space science and technology, we shall consider in this chapter just a few problems, most of which extend or amplify ideas discussed in previous chapters. Calculus is also used in the Appendix.

PROBLEM 1. Until recently it was accepted that there were three possible states in which matter could exist: solid, liquid, and gas. Under conditions that normally prevail on Earth, these are the only states in which matter is found. However, it is now known that if the temperature is very high or the density is very low, a fourth state of matter can exist; it is called *plasma.* A plasma consists of electrons and positively charged ions rather than neutral atoms, and so it has both electric and magnetic fields. (An ion is an atom that has lost one or more of its electrons.) On Earth, plasmas exist, at least temporarily, in lightning, electrical sparks, fluorescent lamps, and in the ionosphere.

In addition to the electromagnetic radiation we sense as heat and light, it is now known that the Sun emits particle radiation having a wide range of energies. The particles (or plasma) appear to come from specific regions on the Sun, some as highly energetic particles which move radially outward into interplanetary space. Some of these highly energetic particles that reach Earth's ionosphere produce auroral displays (the northern lights) and affect shortwave radio transmission by modifying the ionospheric structure.

A lower energy component of the particles is emitted from the Sun on a continuous basis, and these lower energy particles also move away from the Sun in a straight line (radially). The study of this interplanetary plasma, which has been called the solar wind, is of great concern to astronomers and other scientists for several reasons. One is that the Sun is the only star we are close to, and the emission of plasma means that it is very gradually losing matter, an important factor in stellar evolution. Another is that the plasma state of matter is difficult to study on Earth because it is hard to reproduce in the laboratory the conditions of high temperature and low density that exist naturally in the solar atmosphere and in interplanetary space.

A number of space probes and satellites have been used to investigate the properties of the interplanetary plasma. The *Interplanetary Monitoring Platform* (IMP) series of probes from 1963 to the present, the *Orbiting Geophysical Observatory* (OGO) series from 1964 to 1974, the *International Sun-Earth Explorer* (ISEE) satellites from 1977 to the present, and the *Mariner*, *Pioneer*, and *Voyager* deep-space probes have all carried experiments resulting in a series of measurements of flow direction, density, velocity, and electric and magnetic fields of the solar wind.

It has been postulated, on theoretical grounds, that the magnetic field lines of the solar wind coincide with the locus of particles emitted from the Sun, and the experimental findings to date seem to support this hypothesis.

a. Determine the shape of this locus, given that the solar atmosphere from which emission takes place rotates at a constant angular velocity and that particles move outward with constant velocity in the radial direction. Assume the direction of rotation is clockwise.

Solution: An intuitive solution using "time lapse" polar graphing is displayed in Fig. 10.1, and this shows that the locus is an Archimedean spiral. This can be verified analytically using calculus. We are seeking $r = f(\theta)$ such that $dr/dt = V$ (particles emitted with constant radial velocity), and $d\theta/dt = C$ (emissive origin is rotating with constant angular velocity). From the chain rule,

$$\frac{dr}{dt} = \frac{dr}{d\theta}\frac{d\theta}{dt},$$

and substituting from above, we have

$$V = \frac{dr}{d\theta}C, \qquad \text{or} \qquad \frac{dr}{d\theta} = \frac{V}{C},$$

a constant we may call k.

Integrating and choosing the coordinate system so that $f(0) = 0$, we have $r = k\theta$, which is the equation of an Archimedean spiral.

Fig. 10.1

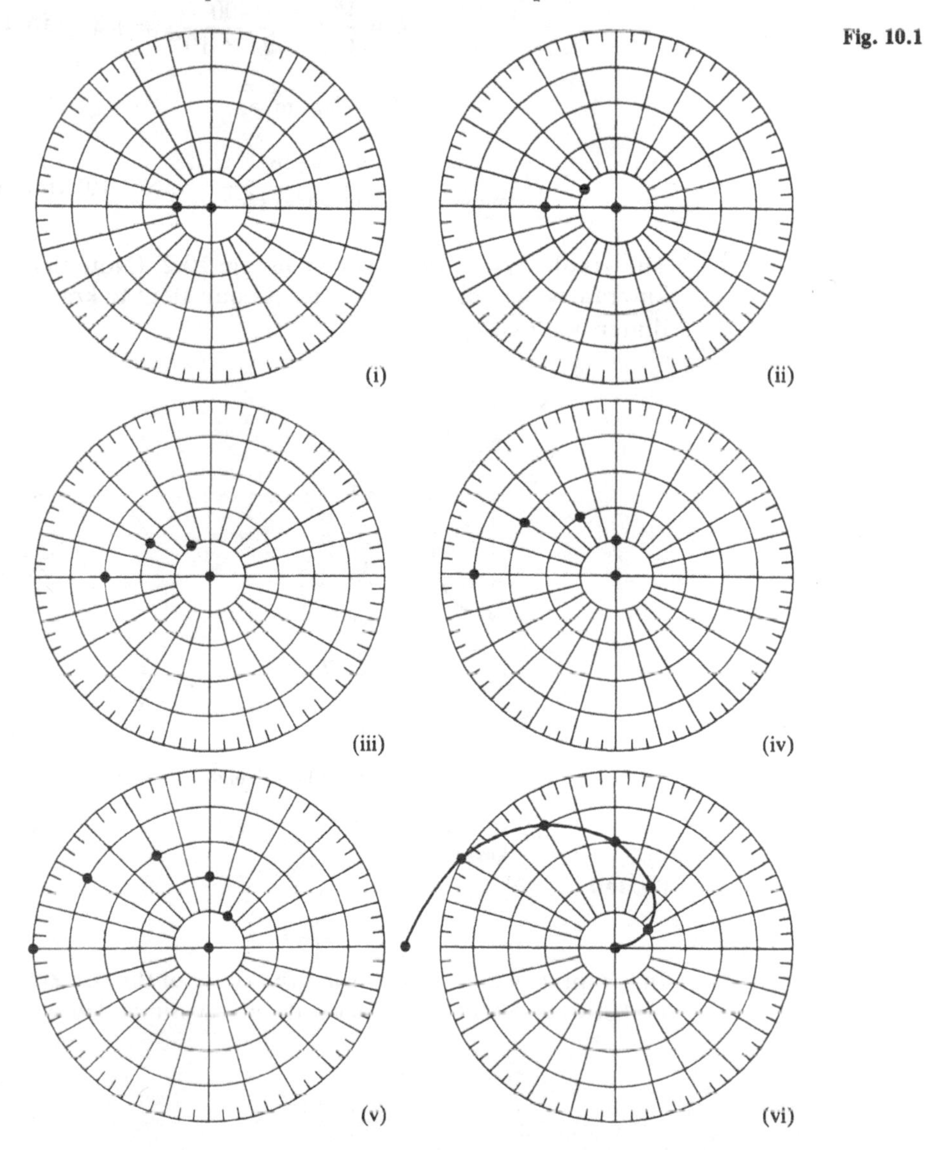

b. It has been observed that the equatorial region of the solar atmosphere rotates at a rate of $C = 2.94 \times 10^{-6}$ radians per second with respect to the distant stars. This is known as the *sidereal rotation rate* and is equivalent to a sidereal rotation period of

$$\frac{2\pi}{2.94 \times 10^{-6}}\ \text{s} = 2.14 \times 10^{6}\ \text{s} = 24.7\ \text{days}.$$

Spacecraft measurements of the solar wind velocity show time variations, with velocity peaks at approximately 25-day intervals as well. For example, *Mariner 2* measured velocities varying from 400 km/s to about 750 km/s at some peaks. Determine k and plot the graphs of the Archimedean spirals for velocities of 400 km/s and 750 km/s.

Solution: For $V = 400$ km/s,

$$k = \frac{V}{C} = \frac{400}{2.94 \times 10^{-6}} = 1.4 \times 10^{8}\ \text{km/rad}.$$

For $V = 750$ km/s,

$$k = \frac{750}{2.94 \times 10^{-6}} = 2.6 \times 10^{8}\ \text{km/rad}.$$

The graphs of $r = k\theta$ are shown in Fig. 10.2. (Note that it is the practice to use km/rad as the unit for k, but this is equivalent to km, since the radian is dimensionless.)

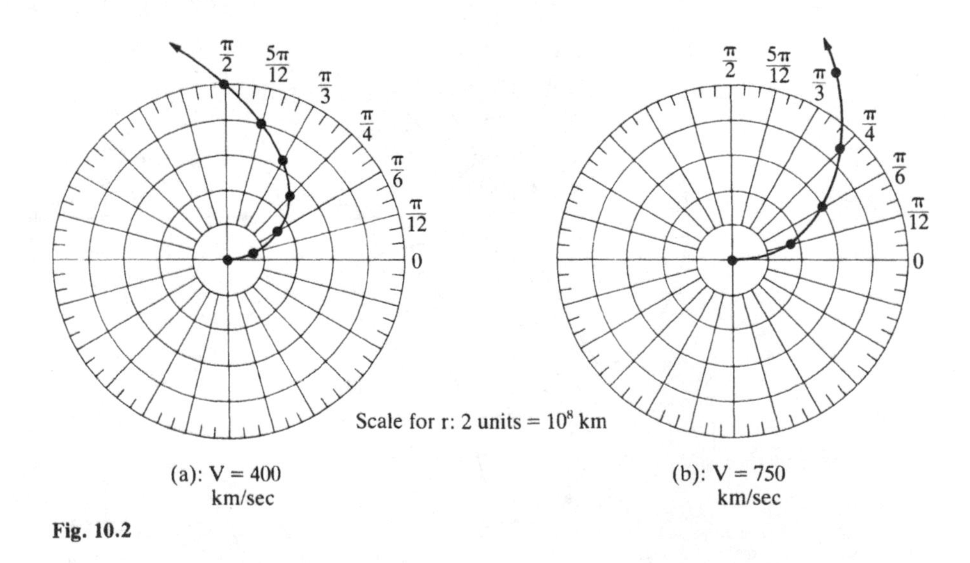

Fig. 10.2

In Chapters 4 and 7, we considered some of the corrections needed to produce undistorted pictures of spacecraft observations. Here is another such correction.

PROBLEM 2. Most satellite photography makes use of scanning techniques. This is illustrated in Fig. 10.3(a), where the scanning is done in the direction orthogonal to the flight path. In performing the scan, a system of mirrors and lenses rotates around an axis parallel to the flight path. Although the scanning system rotates at a constant rate, we can see from Fig. 10.3(b) that the rate at which the scanning beam moves along the ground depends on the angle it makes with the vertical.

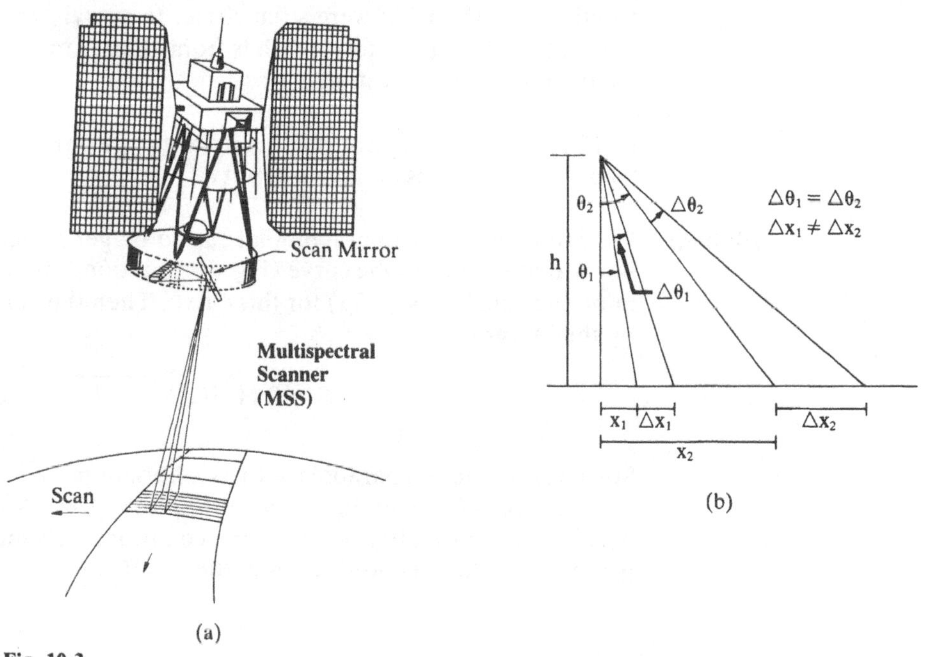

Fig. 10.3

If we imagine that a square on the ground has the pattern shown in Fig. 10.4(a), the result of this variable Earth-scan rate will be the distorted pattern shown in Fig. 10.4(b). In order to produce an undistorted picture, the actual recording of the images must be done at the Earth-scan rate rather than the rotation rate. This *panoramic distortion correction* requires the ability to express the scan rate along the ground, dx/dt in terms of the satellite height, h, the angle θ, and the rotation rate, $d\theta/dt$, of the scanning system, where θ, h, and x are as defined in Fig. 10.3(b). Find such an expression.

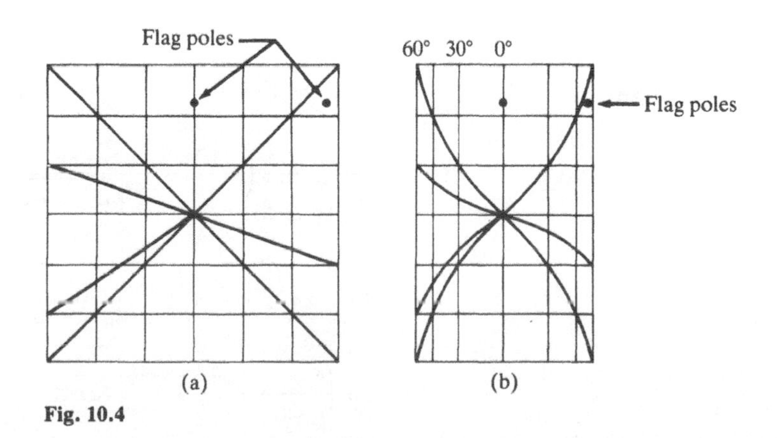

Fig. 10.4

Solution: (This is a straightforward related rate problem.) From Fig. 10.3(b), we see that $\tan\theta = \frac{x}{h}$, so that $x = h\tan\theta$. Differentiating with respect to time,

$$\frac{dx}{dt} = h\sec^2\theta\frac{d\theta}{dt}.$$

PROBLEM 3. In Problem 9a of Chapter 4, we estimated the surface area of an antenna "dish" by treating it as though it were a flat circle. In actual practice, such an antenna is a paraboloidal "cap" whose depth is from 10 percent to 20 percent of its radius. Let us see how good this estimate was.

a. Find an expression for the surface area of a paraboloidal cap that is bounded by a circle of radius r and has depth a.

Solution: We can consider the paraboloidal cap to be generated by revolving the illustrated portion of the parabolic curve (Fig. 10.5) around the x-axis. We must first determine the function $y = f(x)$ for this curve. Then the surface area will be given by the integral

$$S = 2\pi\int_o^a f(x)\sqrt{1 + [f'(x)]^2}\,dx$$

Since this curve is a parabola with axis horizontal, opening to the left, and with vertex at $(a,0)$, its equation has the form $y = b\sqrt{a - x}$, where b must be determined so that $(0,r)$ satisfies the equation. This means $r = b\sqrt{a - 0}$, or $b = r/\sqrt{a}$. The function we need, then, is $f(x) = (r/\sqrt{a})\sqrt{a - x} = r\sqrt{1 - x/a}$.

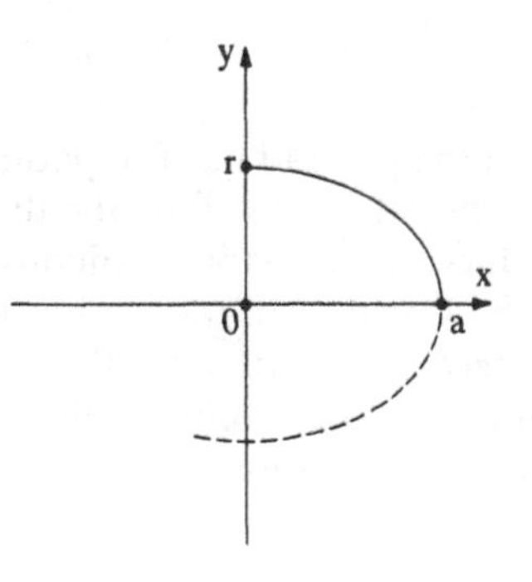

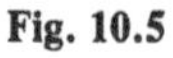
Fig. 10.5

Then

$$f'(x) = \frac{r}{2\sqrt{1 - \frac{x}{a}}}\left(-\frac{1}{a}\right) = \frac{-r}{2\sqrt{a^2 - ax}}$$

$$S = 2\pi\int_0^a r\sqrt{1 - \frac{x}{a}}\sqrt{1 + \frac{r^2}{4a^2 - 4ax}}\,dx$$

$$= 2\pi\int_0^a \frac{r}{2a}\sqrt{r^2 + 4a^2 - 4ax}\,dx$$

This integral may be evaluated by the substitution $u = r^2 + 4a^2 - 4ax$ to produce

$$S = 2\pi \int_{r^2}^{r^2+4a^2} \frac{r}{8a^2}\sqrt{u}\,du = \frac{\pi r}{4a^2}\left[\frac{2}{3}u^{3/2}\right]_{r^2}^{r^2+4a^2}$$

$$= \frac{\pi r}{6a^2}[(r^2 + 4a^2)^{3/2} - r^3]$$

b. Recall that the paraboloidal cap had a radius of 10 meters. If its depth was 1 m, find its surface area (assuming exact numbers) and then find the relative error of the estimate made in Chapter 4.

Solution: For $r = 10$ m and $a = 1$ m,

$$S = \frac{10\pi}{6}[(104)^{3/2} - 10^3] \doteq 317 \text{ m}^2.$$

The estimate, approximating the paraboloidal cap as a circle, was

$$S = \pi r^2 = 100\pi \doteq 314 \text{ m}^2.$$

$$\text{r.e.} = \frac{317 - 314}{317} = 0.0098, \text{ or about 1 percent.}$$

c. Find the relative error for a depth of 2 m.

Solution:

$$S = \frac{10\pi}{24}[(116)^{3/2} - 10^3] \doteq 326 \text{ m}^2$$

$$\text{error} = |326 - 314| \text{ m}^2 = 12 \text{ m}^2$$

$$\text{r.e.} = \frac{12}{326} \doteq 3.8 \text{ percent}$$

PROBLEM 4. In Problem 11 of Chapter 7, we observed that a spacecraft at a distance h from Earth can observe only a portion of Earth's surface.

a. Derive a formula for finding the fraction of the observable area as a function of height above Earth's surface.

Solution: The portion of Earth's surface visible from the spacecraft is shown shaded in Fig. 10.6. Let A_z be the area of the zone with altitude BE. If we set up a rectangular coordinate system with origin at A, and if the coordinates of a point on the arc EC are $(g(y), y)$, then this surface area is found by evaluating the integral

$$A_z = 2\pi \int_{y_B}^{y_E} g(y)\sqrt{1 + [g'(y)]^2}\,dy$$

where $g(y) = x = \sqrt{R^2 - y^2}$ and y_B and y_E are the y-coordinates, respectively, of the points B and E.

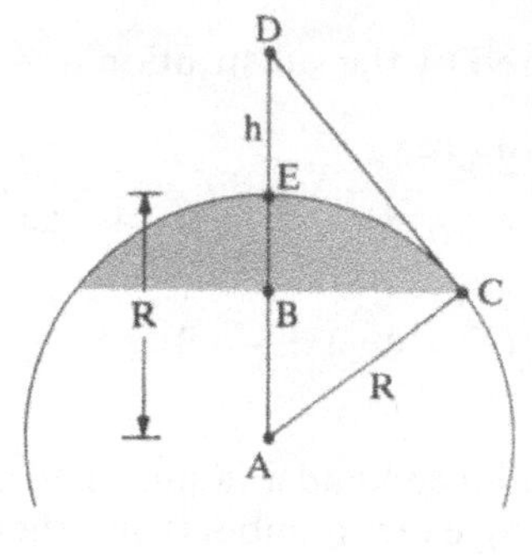

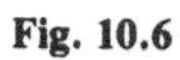
Fig. 10.6

To determine the y-coordinate of B, y_B, we observe that triangles ABC and ACD are similar, so that

$$\frac{AB}{AC} = \frac{AC}{AD}$$

or

$$\frac{y_B}{R} = \frac{R}{R+h},$$

giving

$$y_B = \frac{R^2}{R+h}.$$

We have

$$y_E = R,$$

and

$$g'(y) = \frac{-y}{\sqrt{R^2 - y^2}},$$

so the integral is

$$\begin{aligned} A_z &= 2\pi \int_{\frac{R^2}{R+h}}^{R} \sqrt{R^2 - y^2}\sqrt{\frac{R^2}{R^2 - y^2}}\,dy \\ &= 2\pi \int_{\frac{R^2}{R+h}}^{R} R\,dy \\ &= 2\pi R\left[R - \frac{R^2}{R+h}\right] = \frac{2\pi R^2 h}{R+h} \end{aligned}$$

If we let A_e represent the area of Earth's surface, then $A_e = 4\pi R^2$, so that

$$\frac{A_z}{A_e} = \frac{2\pi R^2 h}{4\pi R^2(R+h)} = \frac{h}{2(R+h)}.$$

b. In April 1983, two members of the Space Shuttle *Challenger* crew, Story Musgrave and Donald Peterson, performed an extravehicular activity (an activity outside the spacecraft) while *Challenger* was at an altitude of 280 km. What fraction of Earth did they see? (Use 6380 km for the Earth's radius.)

Solution: For $h = 280$ and $R = 6380$,

$$\frac{A_z}{A_e} = \frac{280}{2(6380 + 280)} = 0.021, \text{ or } 2.1 \text{ percent.}$$

c. Discuss the manner in which the fraction A_z/A_e varies with the altitude h.

Solution: Intuition suggests that as h increases, the value of A_z/A_e should vary from zero to 1/2. On the surface of Earth, the fraction is zero. As h increases, so does the fraction, and yet it must always be less than 1/2 ; that is, one cannot hope to view more than a hemisphere at any one time. A little algebra bears this out.

$$\frac{A_z}{A_e} = \frac{h}{2(r + h)}$$

is certainly zero when h = 0. Observe that

$$\frac{A_z}{A_e} = \frac{1}{2\left(\frac{r}{h} + 1\right)}.$$

As h increases, the denominator of the right-hand side decreases, which forces the entire fraction A_z/A_e to increase. Furthermore, as $h \to \infty$, $r/h \to 0$, and consequently A_z/A_e approaches $1/[2(1 + 0)] = 1/2$.

d. At what altitude will an astronaut see one-fourth of Earth's surface?

Solution: We must find h such that

$$\frac{1}{4} = \frac{h}{2(6380 + h)}$$

$$4 \cdot h = 2(6380) + 2 \cdot h$$

$$2 \cdot h = 2(6380)$$

$$h = 6380 \text{ km.}$$

e. The first astronauts to travel that far from Earth were the *Apollo 8* crew (Anders, Borman, and Lovell), who orbited the Moon on Christmas Day, 1968. What percent of Earth's surface could these astronauts see as they passed the Moon, a distance of 3.76×10^5 km from Earth?

Solution:

$$\frac{A_z}{A_e} = \frac{3.76 \times 10^5}{2(3.76 \times 10^5 + 6.38 \times 10^3)} = \frac{3.76 \times 10^5}{2(3.82 \times 10^5)}$$

$$= 0.492, \text{ or } 49.2 \text{ percent}$$

PROBLEM 5. The reflective properties of the conic sections were discussed in the final two problems of Chaper 9, where the formula for the slope of the tangent to a parabola was used. Use differentiation to find such a formula for each of the following conics at a point (x_0, y_0).

a. The parabola $y^2 = 4px$

Solution: Differentiation produces

$$2y\frac{dy}{dx} = 4p$$

$$\frac{dy}{dx} = \frac{2p}{y}.$$

So the slope of the tangent at (x_0, y_0) is $\frac{2p}{y_0}$, or, if stated in terms of x_0,

$$\text{slope} = \pm\frac{2p}{\sqrt{4px_0}} = \pm\sqrt{\frac{p}{x_0}}.$$

b. The ellipse $\frac{x^2}{a^2} + \frac{y^2}{b^2} = 1$

Solution: Differentiating this equation,

$$\frac{2x}{a^2} + \frac{2y}{b^2}\frac{dy}{dx} = 0$$

$$\frac{dy}{dx} = -\frac{b^2}{a^2}\frac{x}{y}$$

The slope of the tangent at (x_0, y_0) is then

$$-\frac{b^2}{a^2}\frac{x_0}{y_0}.$$

c. The hyperbola $\frac{x^2}{a^2} - \frac{y^2}{b^2} = 1$

Solution: This is exactly the same as (b) with one sign change, so that the slope of the tangent at (x_0, y_0) is $\frac{b^2}{a^2}\frac{x_0}{y_0}$.

PROBLEM 6. Our final example in this chapter is a result of the spectacular discovery by the *Voyager* space probes that one of Jupiter's moons, Io, is the site of active volcanoes, the first known instance of volcanoes other than those here on Earth. Images returned by the spacecraft have provided measurements which scientists are using to develop and evaluate models by which both the behavior and possible causes of this volcanic activity may be understood.

The starting point is the familiar projectile problem with the valid condition (in contrast to such problems applied to Earth) that air resistance is neglected, since Io has no atmosphere. It is also convenient to begin the modeling process by assuming that the gas and solid particles ejected from the volcano's opening do not affect each other's motion upon ejection, that the opening is circular (roughly), and that all particles start from the same point below the surface with the same initial velocity at all possible escape angles in any direction. These assumptions together have been called the *ballistic model.*

a. Photographs of the plumes of some of Io's active volcanoes show that any vertical cross section through the volcano vent's center has the shape diagrammed in Fig. 10.7, where we have placed the r-axis along Io's surface and the z-axis perpendicular to the surface. If (r, z) is a representative point in such a cross section, find an expression for the escape angle i (or a trigonometric function of i) which will cause an ejected particle to pass through this point (i is measured from the z-axis). Let r_p be the radius of the circular opening, v_0 the initial velocity of the particle, and d the distance below the surface of the point where the particle originates.

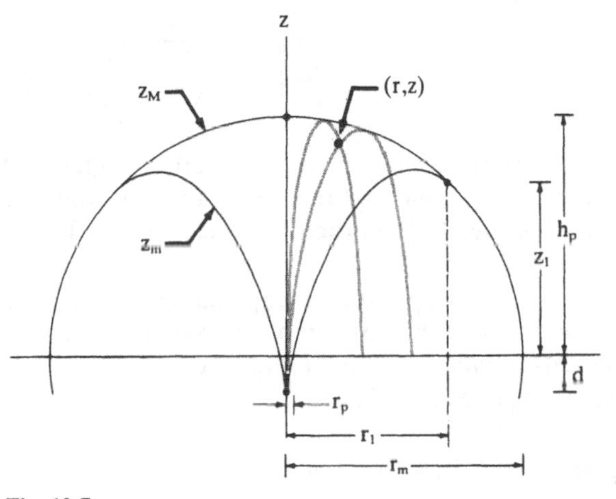

Fig. 10.7

Solution: If g is the value of Io's gravity and t_0 the time at which the particle is ejected, then our model is given mathematically by the following conditions:

$$\frac{d^2z}{dt^2} = -g; \quad \frac{dz}{dt}_{t=t_0} = v_0 \cos i; \quad z(t_0) = -d$$

$$\frac{d^2r}{dt^2} = 0; \quad \frac{dr}{dt}_{t=t_0} = v_0 \sin i; \quad r(t_0) = 0$$

Routine integration and application of the initial conditions result in

(A) $$z(t) = -\frac{1}{2}gt^2 + gt_0t + v_0t\cos i - d - \frac{1}{2}gt_0^2 - v_0t_0\cos i$$

(B) $$r(t) = (v_0 \sin i)(t - t_0)$$

To get an expression for i, we eliminate t by solving equation (B) for t and substituting this expression into equation (A), which gives

(C) $$z = \frac{-gr^2}{2v_0^2}\operatorname{cosec}^2 i + r\cot i - d.$$

Since equation (C) contains $\operatorname{cosec}^2 i$ and $\cot i$, we may use the identity $\operatorname{cosec}^2 i = 1 + \cot^2 i$ to obtain an equation in $\cot i$, namely,

$$\frac{gr^2}{2v_0^2}\cot^2 i - r\cot i + \frac{gr^2}{2v_0^2} + d + z = 0.$$

Applying the quadratic formula,

$$\cot i = r \pm \frac{\sqrt{r^2 - 4\left(\frac{gr^2}{2v_0^2}\right)\left(\frac{gr^2}{2v_0^2} + d + z\right)}}{gr^2/v_0^2},$$

and simplifying,

(D) $$\cot i = \frac{v_0^2}{gr}\left(1 \pm \sqrt{1 - \left(\frac{gr}{v_0^2}\right)^2 - \left(\frac{2g}{v_0^2}\right)(d + z)}\right)$$

We see that there are two possible ejection angles that will bring a particle through a particular point, one on the way up and the other on the way down. This is illustrated in Fig. 10.7, at the point marked (r, z).

b. If i_0 is the largest possible escape angle for the vent, express the height z_m of the lower boundary of the portion of the plume that contains both upward- and downward-moving particles (identified in Fig. 10.7) as a function of r, and then find r_m and r_p of Fig. 10.7.

Solution: From equation (C), with $i = i_0$,

$$z_m = -\frac{gr^2}{2v_0^2}\operatorname{cosec}^2 i_0 + r\cot i_0 - d;$$

r_m and r_p are the values of r in this expression for which $z_m = 0$. Setting $z_m = 0$ and multiplying by

$$-\frac{2v_0^2\sin^2 i_0}{g}$$

produces the equation

$$r^2 - \frac{2v_0^2 \sin i_0 \cos i_0}{g}r + \frac{2v_0^2\, d\sin^2 i_0}{g} = 0.$$

Using the positive sign in the quadratic formula to get the larger r,

$$r_m = \frac{v_0^2}{g}\sin i_0\left[\cos i_0 + \sqrt{\cos^2 i_0 - \frac{2gd}{v_0^2}}\right];$$

and using the negative sign to get the smaller value,

$$r_p = \frac{v_0^2}{g} \sin i_0 \left[\cos i_0 - \sqrt{\cos^2 i_0 - \frac{2gd}{v_0^2}} \right]$$

c. Express the height z_M of the upper boundary of the plume of Fig. 10.7 as a function of r and find an expression for h_p, the maximum height of the plume.

Solution: The upper boundary is the set of points for which the two solutions given by equation (D) of part (a) coalesce; in other words, where the radical vanishes, or

$$1 - \left(\frac{gr}{v_0^2}\right)^2 - \frac{2g\,(d + z_M)}{v_0^2} = 0,$$

so

$$z_M = \frac{v_0^2}{2g} - \frac{gr^2}{2v_0^2} - d.$$

The maximum height h_p occurs when $r = 0$:

$$h_p = \frac{v_0^2}{2g} - d \tag{E}$$

d. Express the coordinates (r_1, z_1) of the point to the right of the z axis, at which the upper and lower boundaries meet, in terms of i_0, v_0, g, and d.

Solution: Since this point is on the upper boundary, we have the radical of equation (D) equal to 0, and since it is on the lower boundary, we have $i = i_0$. From equation (D),

$$\cot i_0 = \frac{v_0^2}{gr_1}, \text{ so } r_1 = \frac{v_0^2}{g} \tan i_0.$$

From our solution to part (c),

$$z_1 = \frac{v_0^2}{2g} - \frac{gr_1^2}{2v_0^2} - d$$

or

$$z_1 = \frac{v_0^2}{2g} - \frac{g}{2v_0^2}\left(\frac{v_0^2}{g} \tan i_0\right)^2 - d$$

$$= \frac{v_0^2}{2g}\,(1 - \tan^2 i_0) - d.$$

e. Measurements of images obtained from the *Voyager 1* imaging system have provided values for r_m, r_p, and h_p. Use the foregoing results to obtain expressions so that i_0, v_0^2/g, and d may be calculated from these measurements.

Solution: From equation (E), $d = (v_0^2/2g) - h_p$. This will give d if we have v_0^2/g, since h_p is known.

From part (b),

$$r_m + r_p = \frac{2v_0^2}{g} \sin i_0 \cos i_0$$

or

$$\frac{v_0^2}{g} = \frac{r_m + r_p}{\sin(2i_0)} = (r_m + r_p) \operatorname{cosec}(2i_0).$$

This will give v_0^2/g if we have i_0, since r_m and r_p are known.

Also from part (b),

$$r_m - r_p = \frac{2v_0^2}{g} \sin i_0 \sqrt{\cos^2 i_0 - \frac{2gd}{v_0^2}}.$$

Since $d = (v_0^2/2g) - h_p$,

$$r_m - r_p = \frac{2v_0^2}{g} \sin i_0 \sqrt{\cos^2 i_0 - 1 + \frac{2gh_p}{v_0^2}}$$

$$= \frac{2v_0^2}{g} \sin i_0 \sqrt{\frac{2gh_p}{v_0^2} - \sin^2 i_0}$$

$$\frac{r_m - r_p}{r_m + r_p} = \frac{\sqrt{\dfrac{2gh_p}{v_0^2} - \sin^2 i_0}}{\cos i_0}.$$

Squaring to eliminate the radical and substituting $\dfrac{g}{v_0^2} = \dfrac{2 \sin i_0 \cos i_0}{r_m + r_p}$,

$$\left(\frac{r_m - r_p}{r_m + r_p}\right)^2 = \frac{\dfrac{2gh_p}{v_0^2} - \sin^2 i_0}{\cos^2 i_0} = \frac{2\left(\dfrac{2 \sin i_0 \cos i_0}{r_m + r_p}\right) h_p - \sin^2 i_0}{\cos^2 i_0}$$

$$= \frac{4h_p}{r_m + r_p} \tan i_0 - \tan^2 i_0,$$

which we can solve for $\tan i_0$ using the quadratic formula:

$$\tan i_0 = \left[\frac{4h_p}{r_m + r_p} \pm \sqrt{\frac{16h_p^2}{(r_m + r_p)^2} - 4\left(\frac{r_m - r_p}{r_m + r_p}\right)^2}\,\right] \Big/ 2$$

Since i_0 appears on the images to be smaller than 45°, the minus sign is chosen, giving

$$\tan i_0 = \frac{2h_p}{r_m + r_p} - \sqrt{4\left(\frac{h_p}{r_m + r_p}\right)^2 - \left(\frac{r_m - r_p}{r_m + r_p}\right)^2}$$

f. *Voyager 1* detected eight volcanic plumes, of which plume 1 and plume 3 were closest in shape to our diagram in Fig. 10.7. Table 10.1 gives the observed measurements of r_m, r_p, and h_p for these plumes. For each, calculate i_0, v_0^2/g, and d from these data, then use the results of part **d.** to predict r_1 and z_1 for these plumes.

Table 10.1
Measurements of Io's Plumes

Plume No.	r_m	r_p	h_p (measured in km)
1	500	17.5	280
3	125	7.5	70

Solution: For plume 1,

$$\tan i_0 = 2\left(\frac{280}{517.5}\right) - \sqrt{4\left(\frac{280}{517.5}\right)^2 - \left(\frac{482.5}{517.5}\right)^2} = 0.533$$

$$\therefore i_0 = 28.05°$$

$$\frac{v_0^2}{g} = (517.5)\,(\text{cosec}\,(56.1°)) = 623 \text{ km}$$

$$d = \frac{v_0^2}{2g} - h_p = \frac{623}{2} - 280 = 32 \text{ km}$$

$$r_1 = \frac{v_0^2}{g} \tan i_0 = (623)\,(0.533) = 332 \text{ km}$$

$$z_1 = \frac{v_0^2}{g}\,(1 - \tan^2 i_0) - d = \frac{623}{2}\left[1 - (0.533)^2\right] - 32 = 191 \text{ km.}$$

For plume 3, the results are as follows:

$$\tan i_0 = 0.4821, \text{ so } i_0 = 25.74°$$

$$\frac{v_0^2}{g} = 169 \text{ km};\ d = 15 \text{ km};\ r_1 = 82 \text{ km};\ z_1 = 50 \text{ km}$$

g. For plume 3, it was also possible to measure r_1. This measurement was 125 km. Compute the relative error of our calculated value for r_1 as a percent.

Solution:

$$\text{Absolute error} = |82 - 125| = 43.$$

$$\text{Relative error} = \frac{43}{125} = 0.34 = 34 \text{ percent.}$$

This error is large enough to demand refinement of the model. The assumption least likely to hold is that ejected gas and particles do not affect each other's motion; it is more probable that the combination of particle sizes and rate of gas flow is such that the particles are carried by the gas into the central portion of the top of the plume and released into ballistic trajectories only on descent.

Although we shall not consider these modifications here, the reader may be interested in knowing that the results of further refinements of the model are consistent with the theoretical proposal that Io's volcanoes are due to tidal effects in its surface generated primarily by another Jovian moon, Europa.

APPENDIX

GRAVITATIONAL FORCES AND THE CONIC SECTION TRAJECTORIES

The elliptical shape of planetary orbits was first asserted by Johannes Kepler on the basis of painstaking observations made by him and by his predecessor, Tycho Brahe. It was Isaac Newton's great achievement to establish mathematically that the inverse square law force of gravitation must produce a trajectory that is one of the conic sections. We present this analysis as an appendix rather than in the calculus chapter (Chapter 10), since the manipulations needed in the development include some complexities that may be unfamiliar to the intended audience.

Before beginning the main problem, we need to establish some properties of first and second derivatives of the vectors. The unit vectors $\vec{i}$, $\vec{j}$, $\vec{u}_r$, $\vec{u}_\theta$ are shown in Fig. A.1. We have

$$\vec{u}_r = \vec{i}\cos\theta + \vec{j}\sin\theta$$

and

$$\vec{u}_\theta = \vec{i}\,(-\sin\theta) + \vec{j}\cos\theta.$$

Differentiating with respect to time,

$$\frac{d\vec{u}_r}{dt} = \vec{i}\,(-\sin\theta)\frac{d\theta}{dt} + \vec{j}\,(\cos\theta)\frac{d\theta}{dt} = \vec{u}_\theta\left(\frac{d\theta}{dt}\right);$$

$$\frac{d\vec{u}_\theta}{dt} = \vec{i}\,(-\cos\theta)\frac{d\theta}{dt} + \vec{j}\,(-\sin\theta)\frac{d\theta}{dt} = -\vec{u}_r\frac{d\theta}{dt};$$

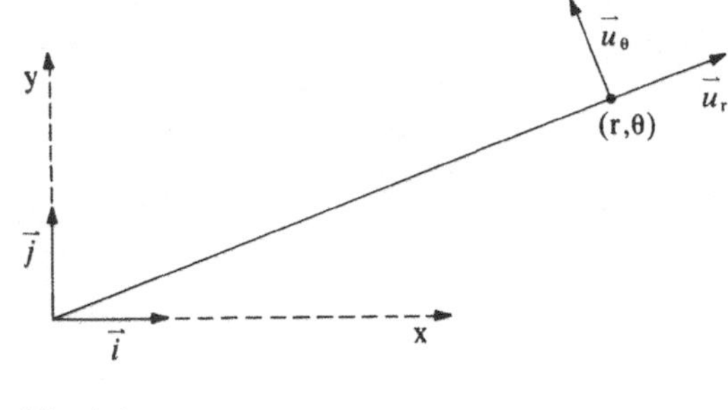

Fig. A.1

In polar coordinates, the position vector is $(r\,\vec{u}_r)$, and so

$$\text{velocity} = \frac{d}{dt}(r\,\vec{u}_r) = \frac{dr}{dt}\vec{u}_r + r\frac{d\vec{u}_r}{dt} = \frac{dr}{dt}\vec{u}_r + r\frac{d\theta}{dt}\vec{u}_\theta;$$

$$\text{acceleration} = \frac{d^2}{dt^2}(r\,\vec{u}_r) = \frac{d}{dt}\left(\frac{dr}{dt}\vec{u}_r + r\frac{d\theta}{dt}\vec{u}_\theta\right)$$

$$= \frac{d^2r}{dt^2}\vec{u}_r + \frac{dr}{dt}\frac{d\vec{u}_r}{dt} + \frac{dr}{dt}\frac{d\theta}{dt}\vec{u}_\theta + r\frac{d^2\theta}{dt^2}\vec{u}_\theta + r\frac{d\theta}{dt}\frac{d\vec{u}_\theta}{dt}$$

$$= \frac{d^2r}{dt^2}\vec{u}_r + 2\frac{dr}{dt}\frac{d\theta}{dt}\vec{u}_\theta + r\frac{d^2\theta}{dt^2}\vec{u}_\theta - r\left(\frac{d\theta}{dt}\right)^2\vec{u}_r$$

$$= \left(\frac{d^2r}{dt^2} - r\left(\frac{d\theta}{dt}\right)^2\right)\vec{u}_r + \left(2\frac{dr}{dt}\frac{d\theta}{dt} + r\frac{d^2\theta}{dt^2}\right)\vec{u}_\theta.$$

We now consider the statement of the law of gravitation: The force on the orbiting body (mass m) is proportional to the product of the masses of the two bodies involved and inversely proportional to the square of the distance between them. In symbols,

$$m\frac{d^2(r\,\vec{u}_r)}{dt^2} = -\frac{GMm}{r^2}\vec{u}_r,$$

where the minus sign expresses the fact that the force of gravitation acts in the direction toward the mass M. Using the result above for acceleration, and canceling the m, we get

$$\left[\frac{d^2r}{dt^2} - r\left(\frac{d\theta}{dt}\right)^2\right]\vec{u}_r + \left[2\frac{dr}{dt}\frac{d\theta}{dt} + r\frac{d^2\theta}{dt^2}\right]\vec{u}_\theta = -\frac{GM}{r^2}\vec{u}_r.$$

Since the $\vec{u}_\theta$ term is missing on the right, we have

$$2\frac{dr}{dt}\frac{d\theta}{dt} + r\frac{d^2\theta}{dt^2} = 0.$$

Kepler's first law, that the radius vector sweeps out equal areas in equal times, can be stated mathematically as $\frac{1}{2}r^2\frac{d\theta}{dt} = A$, a constant. If we differentiate this expression with respect to time, we get

$$\frac{1}{2}\left[2\,r\frac{dr}{dt}\frac{d\theta}{dt} + r^2\frac{d^2\theta}{dt^2}\right] = 0$$

or

$$\frac{1}{2}r\left[2\frac{dr}{dt}\frac{d\theta}{dt} + r\frac{d^2\theta}{dt^2}\right] = 0$$

So we may let the constant $A = r^2\frac{d\theta}{dt}$, and this is equivalent to the fact that the $\vec{u}_\theta$ coefficient vanishes, or Kepler's first law.

Equating the $\vec{u}_r$ terms, $\quad \frac{d^2r}{dt^2} - r\left(\frac{d\theta}{dt}\right)^2 = -\frac{GM}{r^2}.$

Using the substitution $B = GM$, as well as $A = r^2\frac{d\theta}{dt}$, produces the differential equation

$$\text{(1)} \qquad \frac{d^2r}{dt^2} - \frac{A^2}{r^3} + \frac{B}{r^2} = 0.$$

We are seeking r as a function of θ; it turns out to be easier to find $1/r$ as a function of θ. This can be done by letting $w = 1/r$, or equivalently, $r = 1/w$. Then

$$\frac{dr}{dt} = -\frac{1}{w^2}\frac{dw}{dt} = -\frac{1}{w^2}\frac{dw}{d\theta}\frac{d\theta}{dt}$$

$$= -A\frac{dw}{d\theta}, \text{ since } \frac{1}{w^2}\frac{d\theta}{dt} = r^2\frac{d\theta}{dt} = A.$$

Differentiating again,

$$\frac{d^2r}{dt^2} = -A\frac{d}{dt}\left(\frac{dw}{d\theta}\right) = -A\frac{d^2w}{d\theta^2}\frac{d\theta}{dt}$$

$$= -A^2w^2\frac{d^2w}{d\theta^2}$$

We can now rewrite (1) as

$$-A^2w^2\frac{d^2w}{d\theta^2} - A^2w^3 + B^2w^2 = 0$$

or

$$\frac{d^2w}{d\theta^2} + w = \frac{B}{A^2},$$

which has a solution, $w = C\cos(\theta - \theta_0) + \frac{B}{A^2}$.

But this means that

$$r = \frac{1}{w} = \frac{1}{C\cos(\theta - \theta_0) + (B/A^2)} = \frac{A^2/B}{1 + (CA^2/B)\cos(\theta - \theta_0)}.$$

For $A^2/B = ep$, $C = -1/P$, $\theta_0 = 0$, we get $r = ep/(1 - e\cos\theta)$, the equation of the conic section in Chapter 9, Problem 4.

We next derive the "vis-viva," or energy integral, and show that the eccentricity of a conic section trajectory is physically determined by the total energy of the gravitational system. Again, it helps to first establish some properties of the vectors involved. In this context, we will need the square of the velocity vector, $\vec{v}$:

$$v^2 = \vec{v} \cdot \vec{v}, \text{ so } \frac{d(v^2)}{dt} = \frac{d}{dt}(\vec{v} \cdot \vec{v}) = \vec{v} \cdot \frac{d\vec{v}}{dt} + \frac{d\vec{v}}{dt} \cdot \vec{v} = 2\,\vec{v} \cdot \frac{d\vec{v}}{dt}$$

so that

$$\vec{v} \cdot \frac{d\vec{v}}{dt} = \frac{1}{2}\frac{d}{dt}(v^2).$$

The gravitational equation may be written in the form

$$m\frac{d\vec{v}}{dt} = -\frac{GMm}{r^2}\vec{u}_r.$$

Premultiplying by $\vec{v}$, using the dot product, produces

$$m\,\vec{v} \cdot \frac{d\vec{v}}{dt} = -\frac{GMm}{r^2}\vec{v} \cdot \vec{u}_r.$$

Substituting

$$\vec{v} = \frac{dr}{dt}\vec{u}_r + r\frac{d\theta}{dt}\vec{u}_\theta$$

on the right,

$$\vec{v} \cdot \frac{d\vec{v}}{dt} = \frac{1}{2}\frac{d}{dt}(v^2)$$

on the left, and recalling that $\vec{u}_r \cdot \vec{u}_r = 1$ while $\vec{u}_r \cdot \vec{u}_\theta = 0$, we get

$$m\left[\frac{1}{2}\frac{d}{dt}(v^2)\right] = -\frac{GMm}{r^2}\frac{dr}{dt}$$

$$\frac{d}{dt}\left[\frac{1}{2}m\,v^2\right] = GMm\frac{d}{dt}\left(\frac{1}{r}\right).$$

Integrating, we get the energy equation

(2) $$\frac{1}{2}m\,v^2 - \frac{GMm}{r} = E, \text{ a constant.}$$

In physics, the first term on the left, $(1/2)\,m\,v^2$, is the kinetic energy of the system; the second term, $-\frac{GMm}{r}$, is the gravitational potential energy; E, the constant of integration, is the total energy. We may evaluate E by considering a particular point in the orbit. Since $r = ep/(1 - e\cos\theta)$, r attains its minimum value for $\theta = \pi$:

$$r_{\min} = \frac{ep}{1+e}.$$

But

$$\frac{1}{r_{min}} = w_{max},$$

so

$$w_{max} = \frac{1+e}{ep}.$$

Now from the vector expression for velocity,

$$v^2 = \left(\frac{dr}{dt}\right)^2 + \left(r\frac{d\theta}{dt}\right)^2.$$

But when $\theta = \pi$, the velocity is entirely in the $\vec{u}_\theta$ direction, so for $w = w_{max}$, $v^2 = \left(r\frac{d\theta}{dt}\right)^2$ and equation (2) becomes

$$\frac{1}{2}m A^2w^2 - B m w - E = 0,$$

recalling that $A = r^2\frac{d\theta}{dt}$, and $B = GM$. Now, using the quadratic formula with the positive sign,

$$w_{max} = \frac{Bm + \sqrt{(Bm)^2 + 2mA^2E}}{mA^2} = \frac{B}{A^2}\left(1 + \sqrt{1 + \frac{2A^2E}{B^2m}}\right)$$

Equating the two expressions for w_{max}, and recalling that $ep = A^2/B$:

$$\frac{1+e}{ep} = \frac{1+e}{A^2/B} = \frac{B}{A^2}\left(1 + \sqrt{1 + \frac{2A^2E}{B^2m}}\right),$$

and therefore

$$(3) \qquad e = \sqrt{1 + \frac{2A^2E}{B^2m}}.$$

Since $m > 0$, we see that the nature of the trajectory depends on the total energy E:

If $E = 0$, then $e = 1$ and the trajectory is a parabola;

if $E < 0$, then $e < 1$ and the trajectory is an ellipse;

if $E > 0$, then $e > 1$ and the trajectory is a hyperbola.

It is sometimes useful to solve (3) for E, giving

$$
\begin{aligned}
E &= \frac{B^2 m\,(e^2 - 1)}{2A^2} \\
&= \frac{GMm\,(e^2 - 1)}{2\,ep}\,.
\end{aligned}
$$

GLOSSARY

Apogee	The most distant point from Earth reached by a body in an elliptical orbit with Earth at the primary focus.
Attitude of a spacecraft	The orientation of the spacecraft in space, with respect to some chosen coordinate system.
Celestial equator	The projection of the equatorial plane of Earth on the celestial sphere.
Celestial sphere	An imaginary sphere of infinite radius on which celestial objects appear projected.
Center of mass	The point within a body at which all the mass could be located without changing its dynamical behavior.
Conjunction of planets	The position of the planets when they are on the same right ascension circle on the celestial sphere (in other words, when they appear closest together in the sky).
Cosmology	The study of the evolution of the cosmos or universe.
Declination	The analog, on the celestial sphere, of latitude circles on Earth.
Direction cosines	The cosines of the angles made by a vector in space with each of the three positive coordinate axes.
Ecliptic	The path described by the center of the Sun on the geocentric celestial sphere during the course of a year.
Electromagnetic spectrum	Radiation of various wavelengths emitted in the form of waves carrying rapidly varying electric and magnetic fields (light is an example of a portion of the spectrum).
Ephemeris	A list of the successive positions of a celestial object on the geocentric celestial sphere for a series of equally spaced times.
Geocentric	Concentric with Earth.
Iteration	A repetitive mathematical procedure, on an initially chosen trial value, which can produce improved values of a desired quantity.
Jovian	Relating to the planet Jupiter.
Julian day	The number of days, and fraction of a day, measured from noon on 1 January of the year 4713 B.C.
Orbital period	The time it takes for an object to complete one orbit.
Perigee	The closest point to Earth reached by a body in an elliptical orbit with Earth at the primary focus.
Photon	The smallest unit (or "particle") of electromagnetic radiation, carrying one quantum of energy.

Pitch	An angular rotation of an aircraft or spacecraft around an axis through the wings (which has the effect of moving the nose up or down).
Right ascension	The analog, on the celestial sphere, of the longitude circles on Earth.
Right-handed three-dimensional coordinate system	A convention of three basis vectors most simply represented by the thumb, index, and middle fingers of the right hand when extended at right angles to each other; the x-axis is along the thumb, the y-axis along the index finger, and the z-axis along the middle finger.
Roll	An angular rotation of an aircraft or spacecraft around an axis along its length (which has the effect of tipping its wings).
Watt	A unit of power. (In the mks system of units, 1 watt = 1 joule per second.)
Yaw	An angular rotation of an aircraft or spacecraft around an axis perpendicular to its body (which has the effect of moving the nose from side to side).